STUDENT'S SOLUTIONS MANUAL

JAMES LAPP
Colorado Mesa University

ELEMENTARY STATISTICS

THIRTEENTH EDITION

Mario F. Triola

Reproduced by Pearson from electronic files supplied by the author.

Publishing as Pearson, 501 Boylston Street, Boston, MA 02116.

ISBN-13: 978-0-13-446429-9
ISBN-10: 0-13-446429-X

www.pearsonhighered.com

CONTENTS

Chapter 6: Normal Probability Distributions

Chapter 7: Estimating Parameters and Determining Sample Sizes

Chapter 8: Hypothesis Testing

Chapter 9: Inferences from Two Samples

Chapter 10: Correlation and Regression

Chapter 11: Goodness-of-Fit and Contingency Tables

Chapter 12: Analysis of Variance

Chapter 13: Nonparametric Tests

Chapter 14: Statistical Process Control

Chapter 1: Introduction to Statistics

Section 1-1: Statistical and Critical Thinking

1. The respondents are a voluntary response sample or a self-selected sample. Because those with strong interests in the topic are more likely to respond, it is very possible that their responses do not reflect the opinions or behavior of the general population.
3. Statistical significance is indicated when methods of statistics are used to reach a conclusion that a treatment is effective, but common sense might suggest that the treatment does not make enough of a difference to justify its use or to be practical. Yes, it is possible for a study to have statistical significance, but not practical significance.
5. Yes, there does appear to be a potential to create a bias.
7. No, there does not appear to be a potential to create a bias.
9. The sample is a voluntary response sample and has strong potential to be flawed.
11. The sampling method appears to be sound.
13. With only a 1% chance of getting such results with a program that has no effect, the program appears to have statistical significance. Also, because the average loss of 22 pounds does seem substantial, the program appears to also have practical significance.
15. Because there is a 19% chance of getting that many girls by chance, the method appears to lack statistical significance. The result of 1020 girls in 2000 births (51% girls) is above the approximately 50% rate expected by chance, but it does not appear to be high enough to have practical significance. Not many couples would bother with a procedure that raises the likelihood of a girl from 50% to 51%.
17. Yes. Each column of 8 AM and 12 AM temperatures is recorded from the same subject, so each pair is matched.
19. The data can be used to address the issue of whether there is a correlation between body temperatures at 8 AM and at 12 AM. Also, the data can be used to determine whether there are differences between body temperatures at 8 AM and at 12 AM.
21. No. The white blood cell counts measure a different quantity than the red blood cell counts, so their differences are meaningless.
23. No. The National Center for Health Statistics has no reason to collect or present the data in a way that is biased.
25. It is questionable that the sponsor is the Idaho Potato Commission and the favorite vegetable is potatoes.
27. The correlation, or association, between two variables does not mean that one of the variables is the cause of the other. Correlation does not imply causation. Clearly, sour cream consumption is not directly related in any way to motorcycle fatalities.
29. a. 700 adults
    b. 55%
31. a. 559.2 respondents
    b. No. Because the result is a count of respondents among the 1165 engaged or married women who were surveyed, the result must be a whole number.
    c. 559 respondents
    d. 8%
33. Because a reduction of 100% would eliminate all of the size, it is not possible to reduce the size by 100% or more.
35. Because a reduction of 100% would eliminate all plaque, it is not possible to reduce it by more than 100%.
37. The wording of the question is biased and tends to encourage negative responses. The sample size of 20 is too small. Survey respondents are self-selected instead of being randomly selected by the newspaper. If 20 readers respond, the percentages should be multiples of 5, so 87% and 13% are not possible results.

Section 1-2: Types of Data

1. The population consists of all adults in the United States, and the sample is the 2276 adults who were surveyed. Because the value of 33% refers to the sample, it is a statistic.

3. Only part (a) describes discrete data.
5. statistic
7. parameter
9. statistic
11. parameter
13. continuous
15. discrete
17. discrete
19. continuous
21. ordinal
23. nominal
25. interval
27. ordinal
29. The numbers are not counts or measures of anything. They are at the nominal level of measurement, and it makes no sense to compute the average (mean) of them.
31. The temperatures are at the interval level of measurement. Because there is no natural starting point with $0^\circ F$ representing "no heat," ratios such as "twice" make no sense, so it is wrong to say that it is twice as warm at the author's home as it is in Auckland, New Zealand.
33. a. Continuous, because the number of possible values is infinite and not countable.
    b. Discrete, because the number of possible values is finite.
    c. Discrete, because the number of possible values is finite.
    d. Discrete, because the number of possible values is infinite and countable.

Section 1-3: Collecting Sample Data

1. The study is an experiment because subjects were given treatments.
3. The group sample sizes of 547, 550, and 546 are all large so that the researchers could see the effects of the paracetamol treatment.
5. The sample appears to be a convenience sample. By e-mailing the survey to a readily available group of Internet users, it was easy to obtain results. Although there is a real potential for getting a sample group that is not representative of the population, indications of which ear is used for cell phone calls and which hand is dominant do not appear to be factors that would be distorted much by a sample bias.
7. With 717 responses, the response rate is 14%, which does appear to be quite low. In general, a very low response rate creates a serious potential for getting a biased sample that consists of those with a special interest in the topic.
9. systematic
11. random
13. cluster
15. stratified
17. random
19. convenience
21. Observational study. The sample is a convenience sample consisting of subjects who decided themselves to respond. Such voluntary response samples have a high chance of not being representative of the larger population, so the sample may well be biased. The question was posted in an electronic edition of a newspaper, so the sample is biased from the beginning.
23. Experiment. This experiment would create an *extremely* dangerous and illegal situation that has a real potential to result in injury or death. It's difficult enough to drive in New York City while being completely sober.
25. Experiment. The biased sample created by using drivers from New York City cannot be fixed by using a larger sample. The larger sample will still be a biased sample that is not representative of drivers in the United States.
27. Observational study. Respondents who have been convicted of felonies are not likely to respond honestly to the second question. The survey will suffer from a "social desirability bias" because subjects will tend to respond in ways that will be viewed favorably by those conducting the survey.
29. prospective study
31. cross-sectional study
33. matched pairs design
35. completely randomized design
37. a. Not a simple random sample, but it is a random sample.
    b. Simple random sample and also a random sample.
    c. Not a simple random sample and not a random sample.

Quick Quiz

1. No. The numbers do not measure or count anything.
2. nominal
3. continuous
4. quantitative data
5. ratio
6. statistic
7. no
8. observational study
9. The subjects did not know whether they were getting aspirin or the placebo.
10. simple random sample

Review Exercises

1. The survey sponsor has the potential to gain from the results, which raises doubts about the objectivity of the results.
2. a. The sample is a voluntary response sample, so the results are questionable.
   b. statistic
   c. observational study
3. Randomized: Subjects were assigned to the different groups through a process of random selection, whereby they had the same chance of belonging to each group. Double-blind: The subjects did not know which of the three groups they were in, and the people who evaluated results did not know either.
4. No. Correlation does not imply causality.
5. Only part (c) is a simple random sample.
6. Yes. The two questions give the false impression that they are addressing very different issues. Most people would be in favor of defending marriage, so the first question is likely to receive a substantial number of "yes" responses. The second question better describes the issue and subjects are much more likely to have varied responses.
7. a. discrete
   b. ratio
   c. The mailed responses would be a voluntary response sample, so those with strong opinions or greater interest in the topics are more likely to respond. It is very possible that the results do not reflect the true opinions of the population of all full-time college students.
   d. stratified
   e. cluster
8. a. If they have no fat at all, they have 100% less than any other amount with fat, so the 125% figure cannot be correct.
   b. 686
   c. 28%
9. a. interval data; systematic sample
   b. nominal data; stratified sample
   c. ordinal data; convenience sample
10. Because there is a 15% chance of getting the results by chance, those results could easily occur by chance so the method does not appear to have statistical significance. The result of 236 girls in 450 births is a rate of 52.4%, so it is above the 50% rate expected by chance, but it does not appear to be high enough to have practical significance. The procedure does not appear to have either statistical significance or practical significance.

Cumulative Review Exercises

1. The mean is $\frac{3600+1700+4000+3900+3100+3800+2200+3000}{8}=3162.5$ grams. The weights all end with 00, suggesting that all of the weights are rounded to the hundreds place, so that the last two digits are always 00.
2. $0.5^6=0.015625$
3. $\frac{272-176}{6}=16,$ which is an unusually high valuc.
4. $\frac{98.2-98.6}{\frac{0.62}{\sqrt{106}}}=-6.64$
5. $\frac{1.96^2\cdot 0.25}{0.03^2}=1067$
6. $\frac{4000-1700}{4}=575$ grams
7. $\frac{(3600-3162.5)^2}{7}=27,343.75$ grams^2
8. $\sqrt{\frac{(98.4-98.6)^2+(98.6-98.6)^2+(98.8-98.6)^2}{3-1}}=\sqrt{0.04}=0.20$
9. $0.4^8=0.00065536$
10. $9^{11}=31,381,059,609$ (or about $31,381,060,000$)
11. $6^{14}=78,364,164,096$ (or about $78,364,164,000$)
12. $0.3^{12}=0.000000531441$

Chapter 2: Exploring Data with Tables and Graphs

Section 2-1: Frequency Distributions for Organizing and Summarizing Data

1. The table summarizes 50 service times. It is not possible to identify the exact values of all of the original times.

3.

Time (sec)	Relative Frequency
60 – 119	14%
120 – 179	44%
180 – 239	28%
240 – 299	4%
300 – 359	10%

5. Class width: 10
Class midpoints: 24.5, 34.5, 44.5, 54.5, 64.5, 74.5, 84.5
Class boundaries: 19.5, 29.5, 39.5, 49.5, 59.5, 69.5, 79.5, 89.5
Number: 87

7. Class width: 100
Class midpoints: 49.5, 149.5, 249.5, 349.5, 449.5, 549.5, 649.5
Class boundaries: –0.5, 99.5, 199.5, 299.5, 399.5, 499.5, 599.5, 699.5
Number: 153

9. No. The maximum frequency is in the second class instead of being near the middle, so the frequencies below the maximum do not mirror those above the maximum.

11.

Duration (sec)	Frequency
125 – 149	1
150 – 174	0
175 – 199	0
200 – 224	3
225 – 249	34
250 – 274	12

13.

Burger King Lunch Service Times (sec)	Frequency
70 – 109	11
110 – 149	23
150 – 189	7
190 – 229	6
230 – 269	3
230 – 269	6

15. The distribution does not appear to be a normal distribution.

Wendy's Lunch Service Times (sec)	Frequency
70 – 149	25
150 – 229	15
230 – 309	6
310 – 389	3
390 – 469	1

17. Because there are disproportionately more 0s and 5s, it appears that the heights were reported instead of measured. Consequently, it is likely that the results are not very accurate.

x	Frequency
0	9
1	2
2	1
3	3
4	1
5	15
6	2
7	0
8	3
9	1

19. The actresses appear to be younger than the actors.

Age When Oscar Was Won	Relative Frequency (Actresses)	Relative Frequency (Actors)
20 – 29	33.3%	1.1%
30 – 39	39.1%	32.2%
40 – 49	16.1%	41.4%
50 – 59	3.4%	17.2%
60 – 69	5.7%	6.9%
70 – 79	1.1%	1.1%
80 – 89	1.1%	

21.

Age (years) of Best Actress When Oscar Was Won	Cumulative Frequency
Less than 30	29
Less than 40	63
Less than 50	77
Less than 60	80
Less than 70	85
Less than 80	86
Less than 90	87

23. No. The highest relative frequency of 24.8% is not much higher than the others.

Adverse Reaction	Relative Frequency
Headache	23.6%
Hypertension	8.7%
Upper Resp. Tract Infection	24.8%
Nasopharyngitis	21.1%
Diarrhea	21.9%

25. Yes, the frequency distribution appears to be a normal distribution.

Systolic Blood Pressure (mm Hg)	Frequency
80 – 99	11
100 – 119	116
120 – 139	131
140 – 159	34
160 – 179	7
180 – 199	1

27. Yes, the frequency distribution appears to be a normal distribution.

Magnitude	Frequency
1.00 – 1.49	19
1.50 – 1.99	97
2.00 – 2.49	187
2.50 – 2.99	147
3.00 – 3.49	100
3.50 – 3.99	38
4.00 – 4.49	8
4.50 – 4.99	4

29. An outlier can dramatically increase the number of classes.

Weight (lb)	With Outlier	Without Outlier
200 – 219	6	6
220 – 239	5	5
240 – 259	12	12
260 – 279	36	36
280 – 299	87	87
300 – 319	28	28
320 – 339	0	
340 – 359	0	
360 – 379	0	
380 – 399	0	
400 – 419	0	
420 – 439	0	
440 – 459	0	
460 – 479	0	
480 – 499	0	
500 – 519	1	

Section 2-2: Histograms

1. The histogram should be bell-shaped.
3. With a data set that is so small, the true nature of the distribution cannot be seen with a histogram.
5. 40
7. The shape of the graph would not change. The vertical scale would be different, but the relative heights of the bars would be the same.

9. Because it is far from being bell-shaped, the histogram does not appear to depict data from a population with a normal distribution.

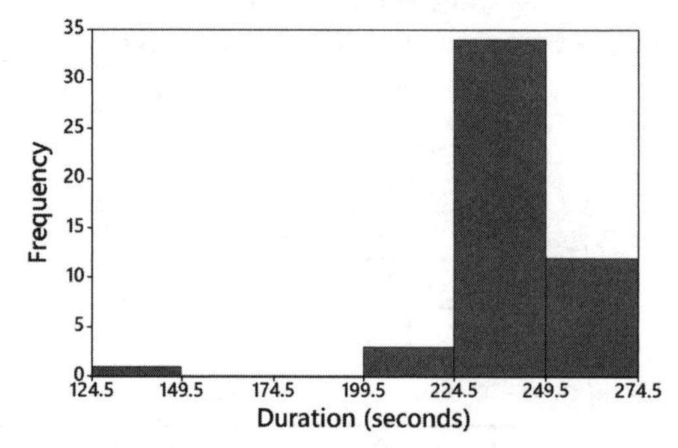

11. The histogram appears to be skewed to the right (or positively skewed).

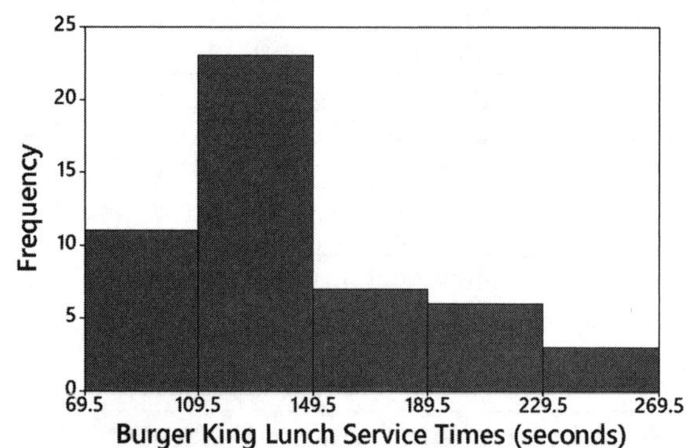

13. The histogram appears to be skewed to the right (or positively skewed).

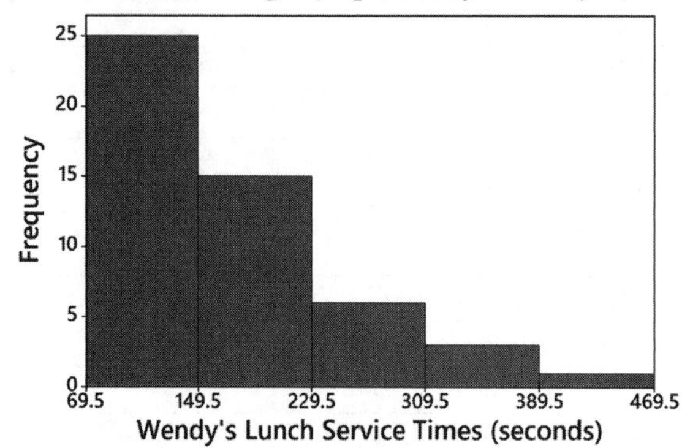

15. The digits 0 and 5 appear to occur more often than the other digits, so it appears that the heights were reported and not actually measured. This suggests that the data might not be very useful.

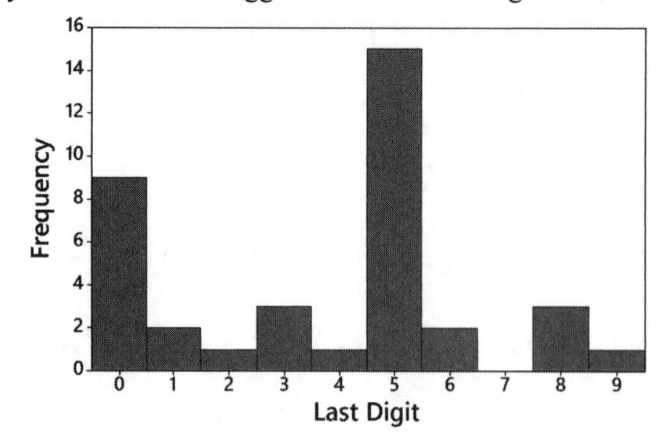

17. The ages of actresses are lower than the ages of actors.

Section 2-3: Graphs That Enlighten and Graphs That Deceive

1. The data set is too small for a graph to reveal important characteristics of the data. With such a small data set, it would be better to simply list the data or place them in a table.
3. No. Graphs should be constructed in a way that is fair and objective. The readers should be allowed to make their own judgments, instead of being manipulated by misleading graphs.
5. The pulse rate of 36 beats per minute appears to be an outlier.

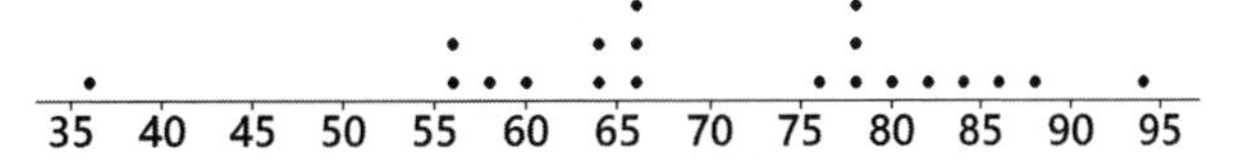

7. The data are arranged in order from lowest to highest, as 36, 56, 56, and so on.

3 | 6
4 |
5 | 668
6 | 044666
7 | 6888
8 | 02468
9 | 4

9. There is a gradual upward trend that appears to be leveling off in recent years. An upward trend would be helpful to women so that their earnings become equal to those of men.

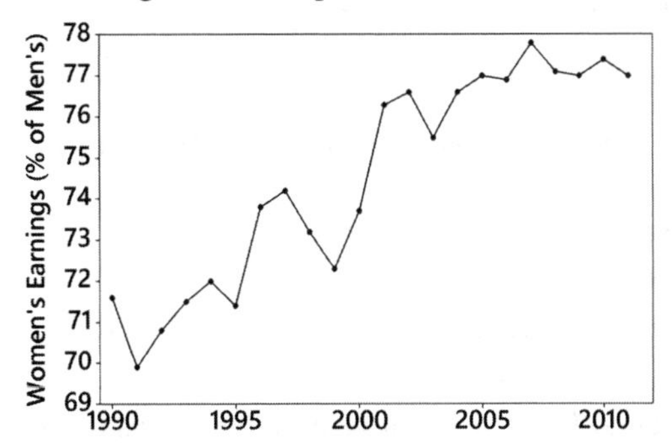

11. Misconduct includes fraud, duplication, and plagiarism, so it does appear to be a major factor.

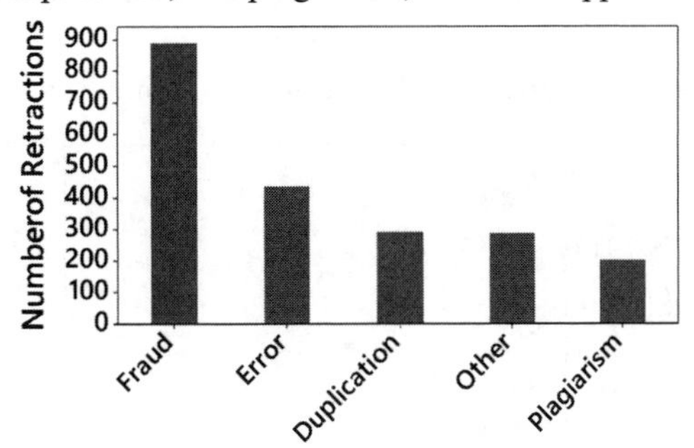

13.

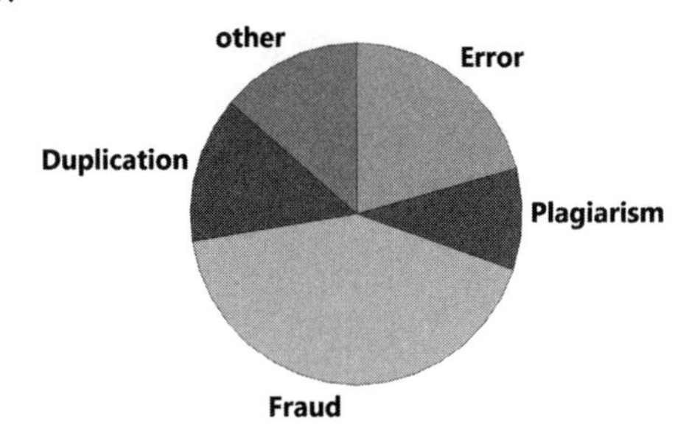

15. The distribution appears to be skewed to the left (or negatively skewed).

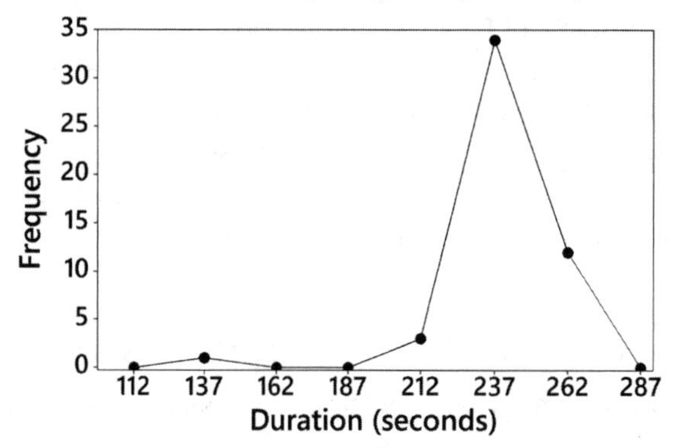

17. Because the vertical scale starts with a frequency of 200 instead of 0, the difference between the "no" and "yes" responses is greatly exaggerated. The graph makes it appear that about five times as many respondents said "no," when the ratio is actually a little less than 2.5 to 1.

19. The two costs are one-dimensional in nature, but the baby bottles are three-dimensional objects. The $4500 cost isn't even twice the $2600 cost, but the baby bottles make it appear that the larger cost is about five times the smaller cost.

21.

96 |
96 | 59
97 | 0001112333444
97 | 55666666788888999
98 | 55556666666666666677777788888889
96 | 001244
96 | 56

Section 2-4: Scatterplots, Correlation, and Regression

1. The term linear refers to a straight line, and r measures how well a scatterplot of the sample paired data fits a straight-line pattern.

3. A scatterplot is a graph of paired (x, y) quantitative data. It helps us by providing a visual image of the data plotted as points, and such an image is helpful in enabling us to see patterns in the data and to recognize that there may be a correlation between the two variables.

5. There does not appear to be a linear correlation between brain volume and IQ score.

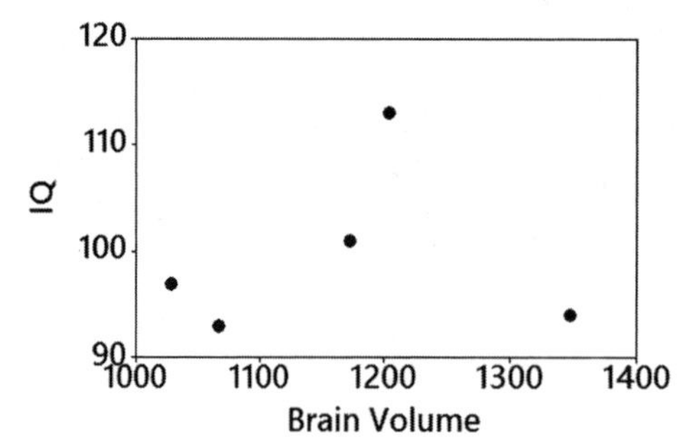

7. There does appear to be a linear correlation between weight and highway fuel consumption.

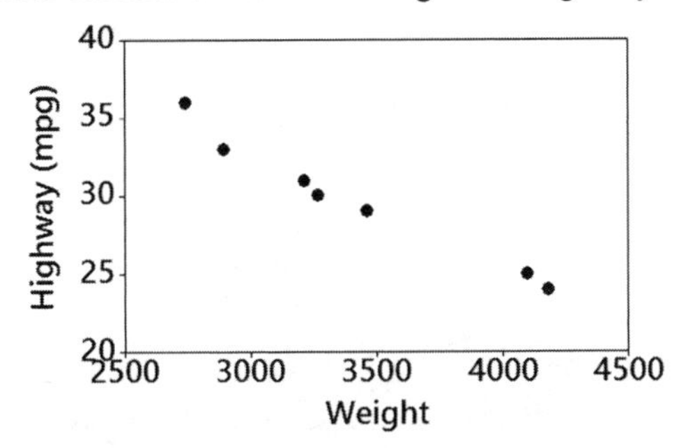

9. With $n = 5$ pairs of data, the critical values are ±0.878. Because $r = 0.127$ is between −0.878 and 0.878, evidence is not sufficient to conclude that there is a linear correlation.

11. With $n = 7$ pairs of data, the critical values are ±0.754. Because $r = 0.987$ is in the left tail region below −0.754, there are sufficient data to conclude that there is a linear correlation.

13. Because the *P*-value is not small (such as 0.05 or less), there is a high chance (83.9% chance) of getting the sample results when there is no correlation, so evidence is not sufficient to conclude that there is a linear correlation.

15. Because the *P*-value is small (such as 0.05 or less), there is a small chance of getting the sample results when there is no correlation, so there is sufficient evidence to conclude that there is a linear correlation.

Quick Quiz

1. Class width: 3. It is not possible to identify the original data values.
2. Class boundaries: 17.5 and 20.5
   Class limits: 18 and 20.
3. 40
4. 19 and 19
5. pareto chart
6. histogram
7. scatterplot
8. No, the term "normal distribution" has a different meaning than the term "normal" that is used in ordinary speech. A normal distribution has a bell shape, but the randomly selected lottery digits will have a uniform or flat shape.
9. variation
10. The bars of the histogram start relatively low, increase to some maximum, and then decrease. Also, the histogram is symmetric, with the left half being roughly a mirror image of the right half.

Review Exercises

1.

Temperature (°F)	Frequency
97.0–97.4	2
97.5–97.9	4
98.0–98.4	7
98.5–98.9	5
99.0–99.4	2

2. Yes, the data appear to be from a population with a normal distribution because the bars start low and reach a maximum, then decrease, and the left half of the histogram is approximately a mirror image of the right half.

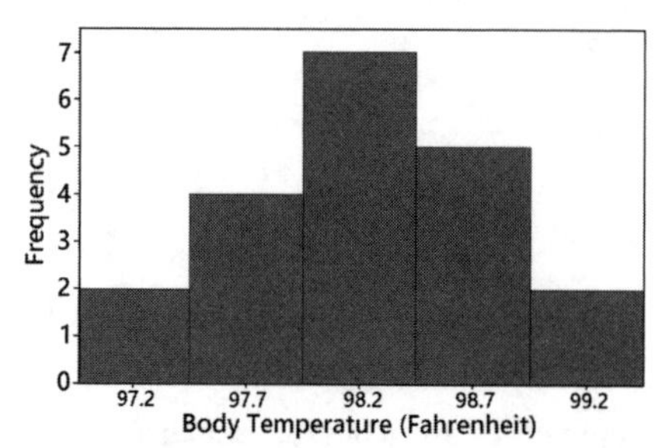

3. The distribution is closer to being a normal distribution than the others.

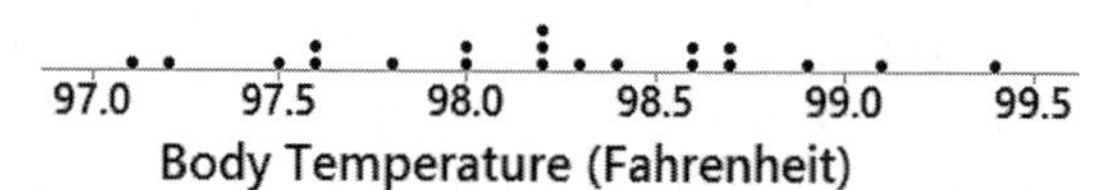

4. There are no outliers.

```
97.|125668
98.|002223466779
99.|14
```

5. No. There is no pattern suggesting that there is a relationship.

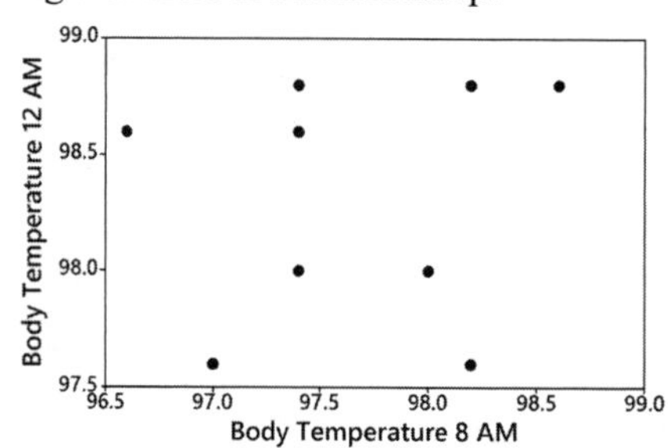

6. a. time-series graph
   b. scatterplot
   c. pareto chart

7. A pie chart wastes ink on components that are not data; pie charts lack an appropriate scale; pie charts don't show relative sizes of different components as well as some other graphs, such as a Pareto chart.

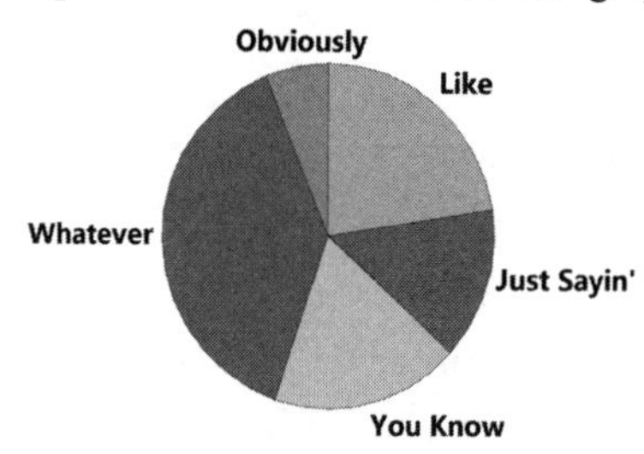

8. The Pareto chart does a better job. It draws attention to the most annoying words or phrases and shows the relative sizes of the different categories.

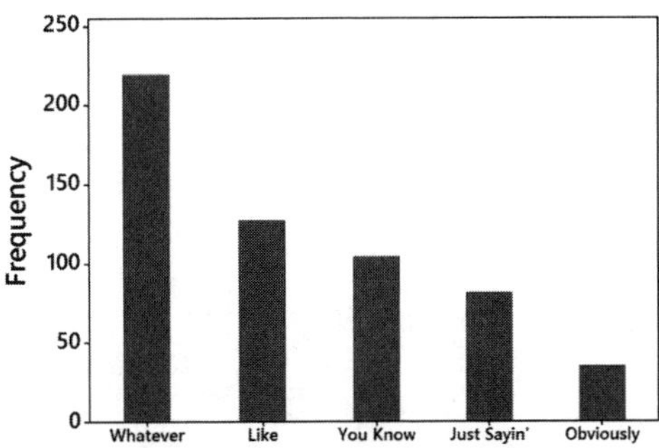

Cumulative Review Exercises

1.

Total Hours	Frequency
235–239	4
240–244	3
245–249	9
250–254	8
255–259	3
260–264	3

2. a. 235 hours and 239 hours
   b. 234.5 hours and 239.5 hours
   c. 237 hours
3. The distribution is closer to being a normal distribution than the others.

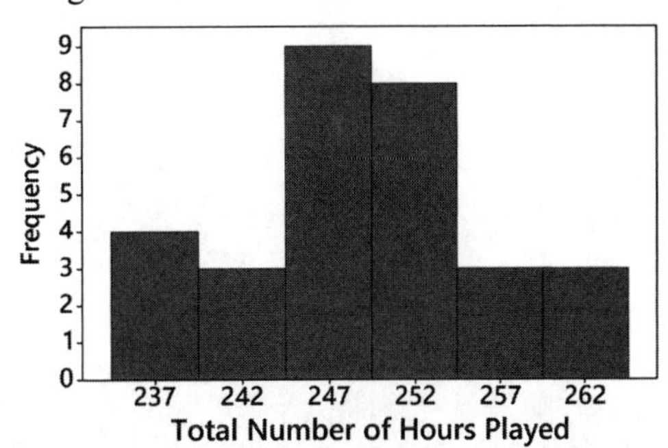

4. Start the vertical scale at a frequency of 2 instead of the frequency of 0.
5. Looking at the stemplot sideways, we can see that the distribution approximates a normal distribution.

```
23 | 6789
24 | 112555677889
25 | 00012233888
26 | 024
```

6. a. continuous
   b. quantitative
   c. ratio
   d. convenience sample
   e. sample

Chapter 3: Describing, Exploring, and Comparing Data

Section 3-1: Measures of Center

1 The term *average* is not used in statistics. The term *mean* should be used for the result obtained by adding all of the sample values and dividing the total by the number of sample values.

3. They use different approaches for providing a value (or values) of the center or middle of the sorted list of data.

5. The mean is $\bar{x}=\dfrac{7+19+20+25+55+60+81+82+89+91+99}{11}=57.1.$

   The median is 60.
   There is no mode.

   The midrange is $\dfrac{7+99}{11}=53.$

   The jersey numbers are nominal data that are just replacements for names, and they do not measure or count anything, so the resulting statistics are meaningless.

7. The mean is $\bar{x}=\dfrac{150+150+150+150+160+160+165+185+200+250}{10}=172$, or \$172.0 million.

   The median is $\dfrac{160+160}{2}=160$, or \$160 million.

   The mode is \$150 million

   The midrange is $\dfrac{150+250}{2}=200$, or \$200 million.

   Apart from the fact that all other celebrities have amounts of net worth lower than those given, nothing meaningful can be known about the population of net worth of all celebrities. The numbers all end in 0 or 5, and they appear to be rounded estimates.

9. The mean is $\bar{x}=\dfrac{2+4+5+6+7+7+8+8+8+8+9+9+12+15}{14}=7.7$ hurricanes.

   The median is $\dfrac{8+8}{2}=8.0$ hurricanes.

   The mode is 8 hurricanes.

   The midrange is $\dfrac{2+15}{2}=6.5$ hurricanes.

   The data are time-series data, but the measures of center do not reveal anything about a trend consisting of a pattern of change over time.

11. The mean is $\bar{x}=\dfrac{950+1150+1200+1400+\cdots+1600+1600+1750+1800}{12}=1454.2$, or \$1454.20.

    The median is $\dfrac{1500+1500}{2}=1500.0$, or \$1500.00.

    The mode is \$1500.

    The midrange is $\dfrac{950+1800}{2}=1375.0$, or \$1375.00.

    The sample consists of "best buy" TVs, so it is not a random sample and is not likely to be representative of the population. The lowest price is a relevant statistic for someone planning to buy one of the TVs.

13. The mean is $\overline{x} = \dfrac{0+0+0+\cdots+34+36+38+41+41+41+\cdots+53+54+55}{20} = 32.0$ mg.

    The median is $\dfrac{38+41}{2} = 39.5$ mg.

    The mode is 0 mg.

    The midrange is $\dfrac{0+55}{2} = 27.5$ mg.

    Americans consume some brands much more often than others, but the 20 brands are all weighted equally in the calculations, so the statistics are not necessarily representative of the population of all cans of the same 20 brands consumed by Americans.

15. The mean is $\overline{x} = \dfrac{8.6+9.1+9.1+9.1+9.3+9.3+9.4+9.8+9.9+10+10.4}{11} = 9.45$ in.

    The median is 9.30 in.
    The mode is 9.1 in.

    The midrange is $\dfrac{8.6+10.4}{2} = 9.50$ in.

    Because the measurements were made in 1988, they are not necessarily representative of the current population of all Army women.

17. The mean is $\overline{x} = \dfrac{39+50+50+50+\cdots+175+200+209+\cdots+500+500+1500+2500}{25} = 365.3$, or \$365.30.

    The median is 200.0, or \$200.00.
    The median is 500, or \$500.00.

    The midrange is $\dfrac{39+2500}{2} = 1269.5$, or \$1269.5.

    The amounts of \$1500 and \$2500 appear to be outliers.

19. The mean is $\overline{x} = \dfrac{0+0+0+0+0+0+\cdots+9+10+10+20+40+50}{50} = 2.8$ cigarettes.

    The median is 0 cigarettes.
    The mode is 0 cigarettes.

    The midrange is $\dfrac{0+50}{2} = 25.0$ cigarettes.

    Because the selected subjects report the number of cigarettes smoked, it is very possible that the data are not at all accurate. And what about that person who smokes 50 cigarettes (or 2.5 packs) a day? What are they thinking?

21. Systolic:

    The mean is $\overline{x} = \dfrac{96+116+118+120+122+126+128+136+156+158}{10} = 127.6$ mm Hg,

    The median is $\dfrac{122+126}{2} = 124.0$ mm Hg.

    Diastolic:

    The mean is $\overline{x} = \dfrac{52+58+64+72+74+76+80+82+88+90}{10} = 73.6$ mm Hg.

    The median is $\dfrac{74+76}{2} = 75.0$ mm Hg.

    Given that systolic and diastolic blood pressures measure different characteristics, a comparison of the measures of center doesn't make sense. Because the data are matched, it would make more sense to investigate whether there is an association or *correlation* between systolic blood pressure measurements and diastolic blood pressure measurements.

23. Males:

The mean is $\bar{x} = \dfrac{54+56+58+\cdots+66+66+72+\cdots+80+86+96}{15} = 69.5$, beats per minute.

The median is 66.0 beats per minute.

Females:

The mean is $\bar{x} = \dfrac{64+68+70+\cdots+82+84+86+\cdots+90+90+94}{15} = 82.1$, beats per minute.

The median is 84.0 beats per minute.

The pulse rates of males appear to be lower than those of females.

25. The mean is $\bar{x} = 0.8$ and the median is 1.0. Ten of the tornadoes have missing F-scale measurements.

27. The mean is $\bar{x} = 98.20°\text{F}$ and the median is $98.40°\text{F}$. These results suggest that the mean is less than 98.6°F.

29. The mean is $\bar{x} = \dfrac{29(24.5)+34(34.5)+14(44.5)+3(54.5)+5(64.5)+1(74.5)+1(84.5)}{29+34+14+3+5+1+1} = 36.2$ years. This result is the same as the mean of 36.2 years found by using the original list of data values.

31. The mean is $\bar{x} = \dfrac{1(49.5)+51(149.5)+90(249.5)+10(349.5)+0(449.5)+0(549.5)+1(649.5)}{1+51+90+10+0+0+1}$

$= 224.0\,(1000 \text{ cells}/\mu\text{L})$. The mean from the frequency distribution is quite close to the mean of

$= 224.3\,(1000 \text{ cells}/\mu\text{L})$. obtained by using the original list of values.

33. The mean is $\bar{x} = \dfrac{3(4)+3(2)+3(3)+4(4)+1(1)}{3+3+3+4+1} = 3.14$, so the student made the dean's list.

35. a. The missing value is $5(78.0)-82-78-56-84 = 90$ beats per minute.

b. $n-1$

37. 504 lb is an outlier. For the original data, the median is 285.5 lb and the mean is 294.4 lb. The 10% trimmed mean is 285.4 lb and the 20% trimmed mean is 285.8 lb. The median, 10% trimmed mean, and 20% trimmed mean are all quite close, but the untrimmed mean of 294.4 lb differs from them because it is strongly affected by the inclusion of the outlier.

39. The geometric mean is $\sqrt[6]{(1.05154)(1.02730)(1.00488)(1.00319)(1.00313)(1.00268)} = 1.015289767$, or 1.015290 when rounded, so the growth rate is 1.5290%.

41. The median is $125+(50)\left(\left(\dfrac{50+1}{2}-(11+1)\right)\Big/2\right) = 153.125$ seconds, which is rounded to 153.1 seconds.

This value differs by 2.6 seconds from the median of 150.5 seconds found by using the original list of service times. The value of 150.5 seconds is better because it is based on the original data and does not involve interpolation.

Section 3-2: Measures of Variation

1. $s \approx \dfrac{1439-963}{4} = 119.0 \text{ cm}^3$, which is quite close to the exact value of the standard deviation of 124.9 cm^3.

3. $(20.0414 \text{ kg})^2 = 401.6577 \text{ kg}^2$

5. The range is $99-7 = 92.0$.

The variance is $s^2 = \dfrac{(7-57.1)^2+(19-57.1)^2+\cdots+(91-57.1)^2+(99-57.1)^2}{11-1} = 1149.5$.

The standard deviation is $s = \sqrt{1149.5} = 33.9$.

The jersey numbers are nominal data that are just replacements for names, and they do not measure or count anything, so the resulting statistics are meaningless.

7. The range is $250-150=100$ million dollars.

   The variance is $s^2=\frac{(150-172)^2+(150-172)^2+\cdots+(200-172)^2+(250-172)^2}{10-1}=1034.4$ (million dollars)^2.

   The standard deviation is $s=\sqrt{1034.4}=32.0$ million dollars.

   Because the data are from celebrities with the highest net worth, the measures of variation are not at all typical for all celebrities.

9. The range is $15-2=13$ hurricanes.

   The variance is $s^2=\frac{14(966)-(108)^2}{14(14-1)}=10.2$ hurricanes^2.

   The standard deviation is $s=\sqrt{10.2}=3.2$ hurricanes.

   Data are time-series data, but the measures of variation do not reveal anything about a trend consisting of a pattern of change over time.

11. The range is $1800-950=\$850$.

    The variance is $s^2=\frac{12(26{,}047{,}500)-(17{,}450)^2}{12(12-1)}=61{,}117.4$ (dollars)^2.

    The standard deviation is $s=\sqrt{61{,}117.4}=\$247.2$.

    The sample consists of "best buy" TVs, so it is not a random sample and is not likely to be representative of the population. The measures of variation are not likely to be typical of all TVs that are 60 inches or larger.

13. The range is $55-0=55.0$ mg.

    The variance is $s^2=\frac{20(29{,}045)-(651)^2}{20(20-1)}=413.4$ mg^2.

    The standard deviation is $s^2=\sqrt{413.4}=20.3$ mg.

    Americans consume some brands much more often than others, but the 20 brands are all weighted equally in the calculations, so the statistics are not necessarily representative of the population of all cans of the same 20 brands consumed by Americans.

15. The range is $10.4-8.6=1.80$ in.

    The variance is $s^2=\frac{11(985.94)-(107)^2}{11(11-1)}=0.27$ in.^{2}

    The standard deviation is $s=\sqrt{0.27}=0.52$ in.

    Because the measurements were made in 1988, they are not necessarily representative of the current population of all Army women.

17. The range is $2500-39=\$2461.0$.

    The variance is $s^2=\frac{25(10{,}306{,}077)-(9133)^2}{25(25-1)}=290{,}400.4$ (dollars)^2.

    The standard deviation is $s=\sqrt{290{,}400.4}=\$538.9$.

    The amounts of \$1500 and \$2500 appear to be outliers, and it is likely that they have a large effect on the measures of variation.

19. The range is $50-0=50.0$ cigarettes.

    The variance is $s^2=\frac{50(4781)-(139)^2}{50(50-1)}=89.7$ (cigarettes)^2.

    The standard deviation is $s=\sqrt{89.7}=9.5$ cigarettes.

    Because the selected subjects report the number of cigarettes smoked, it is very possible that the data are not at all accurate, so the results might not reflect the actual smoking behavior of California adults.

21. Systolic: $\bar{x} = 127.6$, $s = 18.6$; The coefficient of variation is $\frac{18.6}{127.6} \cdot 100\% = 14.6\%$.

    Diastolic: $\bar{x} = 73.6$, $s = 12.5$; The coefficient of variation is $\frac{12.5}{73.6} \cdot 100\% = 16.9\%$.

    The variation is roughly about the same.

23. Male: $\bar{x} = 69.5$, $s = 11.3$; The coefficient of variation is $\frac{11.3}{69.5} \cdot 100\% = 16.2\%$.

    Female: $\bar{x} = 82.1$, $s = 9.2$; The coefficient of variation is $\frac{9.2}{82.1} \cdot 100\% = 11.2\%$.

    Pulse rates of males appear to vary more than pulse rates of females.

25. Range = 4.0, $s^2 = 0.9$, $s = 0.9$

27. Range = 3.10°F, $s^2 = 0.39\ (°\text{F})^2$, $s = 0.62°\text{F}$

29. The rule of thumb standard deviation is $s \approx \frac{4-0}{4} = 1.0$, which is very close to $s = 0.9$ found by using all of the data.

31. The rule of thumb standard deviation is $s \approx \frac{99.6-96.5}{4} = 0.78°\text{F}$, which is not substantially different from $s = 0.62°\text{F}$ found by using all of the data.

33. Significantly low values are less than or equal to $74.0 - 2(12.5) = 49.0$ beats per minute, and significantly high values are greater than or equal to $74.0 + 2(12.5) = 99.0$ beats per minute. A pulse rate of 44 beats per minute is significantly low.

35. Significantly low values are less than or equal to $77.32 - 2(1.29) = 24.74$ cm, and significantly high values are greater than or equal to $77.32 + 2(1.29) = 29.90$ cm. A foot length of 30 cm is significantly high.

37. $s = \sqrt{\frac{87\left(29 \cdot 24.5^2 + \cdots + 1 \cdot 84.5^2\right) - \left(29 \cdot 24.5 + \cdots + 1 \cdot 84.5\right)^2}{87(87-1)}} = 12.7$ years, which differs from the exact value of 11.5 years by a somewhat large amount.

39. $s = \sqrt{\frac{103\left(1 \cdot 49.5^2 + \cdots + 1 \cdot 649.5^2\right) - \left(1 \cdot 49.5 + \cdots + 1 \cdot 649.5\right)^2}{103(103-1)}} = 68.4$, which is somewhat far from the exact value of 59.5.

41. a. The empirical rule states that approximately 95% of women should fall between two standard deviations of the mean.

    b. Since $\frac{189.7-255.1}{65.4} = -1$ and $\frac{320.5-255.1}{65.4} = 1$, the empirical rule states that approximately 68% of women should fall between one standard deviation of the mean.

43. At least $1 - 1/3^2 = 89\%$ of women have platelet counts within 3 standard deviations of the mean. The minimum count is $255.1 - 3(65.4) = 58.9$ and the maximum count is $255.1 + 3(65.4) = 451.3$.

45. a. $\mu = \frac{9+10+20}{3} = 13$ cigarettes and $\sigma^2 = \frac{(9-13)^2 + (10-13)^2 + (20-13)^2}{3} = 24.7$ cigarettes^2

    b. The nine possible samples of two values are the following: {(9, 9), (9, 10), (9, 20), (10, 9), (10, 10), (10, 20), (20, 9), (20, 10), (20,20)} which have the following corresponding sample variances: {0, 0.5, 60.5, 0.5, 0, 50, 60.5, 50, 0}, that have a mean of $s^2 = 2.47$ cigarettes^2.

45. (continued)
    c. The population variances of the nine samples above are {0, 0.25, 30.25, 0.25, 0, 25, 30.25, 25, 0} that have a mean of $s^2 = 12.3\text{ cigarettes}^2$.
    d. Part (b), because repeated samples result in variances that target the same value (24.7 cigarettes^2) as the population variance. Use division by $n-1$.
    e. No. The mean of the sample variances (24.7 cigarettes^2) equals the population variance (24.7 cigarettes^2), but the mean of the sample standard deviations (3.5 cigarettes) does not equal the population standard deviation (5.0 cigarettes).

Section 3-3: Measures of Relative Standing and Boxplots

1. James' height is 4.07 standard deviations above the mean.
3. The bottom boxplot represents weights of women, because it depicts weights that are generally lower.
5. a. The difference is $77.8 - 17.60 = 60.20$ Mbps.
   b. $\dfrac{60.20}{16.02} = 3.76$ standard deviations
   c. $z = 3.76$
   d. The data speed of 77.8 Mbps is significantly high.
7. a. The difference is $36 - 74.0 = -38$ beats per minute.
   b. $\dfrac{38}{12.5} = 3.04$ standard deviation
   c. $z = -3.04$
   d. The pulse rate of 36 beats per minute is significantly low.
9. Significantly low scores are less than or equal to $21.1 - 2(5.1) = 10.9$, and significantly high scores are greater than or equal to $21.1 + 2(5.1) = 31.3$.
11. Significantly low weights are less than or equal to $5.63930 - 2(0.06194) = 5.51542$ g, and significantly high weights are greater than or equal to $5.63930 + 2(0.06194) = 5.76318$ g.
13. The tallest man's z score is $z = \dfrac{251 - 174.12}{7.10} = 10.83$ and the shortest man's z score is $z = \dfrac{54.6 - 174.12}{7.10} = -16.83$. Chandra Bahadur Dangi has the more extreme height because his z score of -16.83 is farther from the mean than the z score of 10.83 for Sultan Kosen
15. The male has a more extreme birth weight because his z score is $z = \dfrac{1500 - 3272.8}{660.2} = -2.19$, which is a lower number than the z score of $z = \dfrac{1500 - 3037.1}{706.3} = -2.18$ for the female.
17. For 2.4 Mbps, $\dfrac{29}{50} \cdot 100 = 58$, so it is the 58th percentile.
19. For 0.7 Mbps, $\dfrac{17}{50} \cdot 100 = 34$, so it is the 34th percentile.
21. $L = \dfrac{60 \cdot 50}{100} = 30$, so $P_{60} = \dfrac{2.4 + 2.5}{2} = 2.45$ Mbps (Tech: Excel: 2.44 Mbps)
23. $L = \dfrac{75 \cdot 50}{100} = 37.5$, so $Q_3 = P_{75} = 3.8$ Mbps (Tech: Minitab: 3.85 Mbps; Excel: 3.75 Mbps)
25. $L = \dfrac{50 \cdot 50}{100} = 25$, so $P_{10} = \dfrac{1.6 + 1.6}{2} = 1.6$ Mbps

27. $L = \frac{25 \cdot 50}{100} = 12.5$, so $P_{25} = 0.5$ Mbps

29. The five number summary is 2, 6.1, 7.0, 8.0, 10.

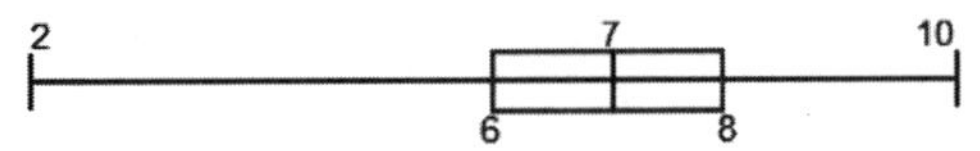

31. The five number summary is 128 mBq, 140.0 mBq, 150.0 mBq, 158.5 mBq, 172 mBq (Tech: Minitab yields $Q_1 = 139.0$ mBq and $Q_3 = 159.75$ mBq. Excel yields $Q_1 = 141.0$ mBq and $Q_3 = 157.25$ mBq.)

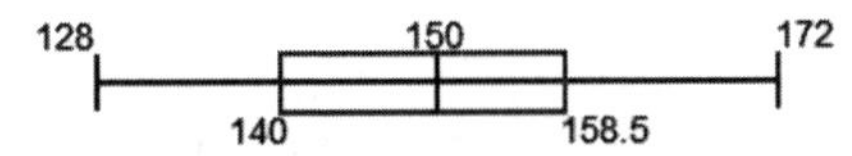

33. The top boxplot represents males. Males appear to have slightly lower pulse rates than females.

Male Pulse

Female Pulse

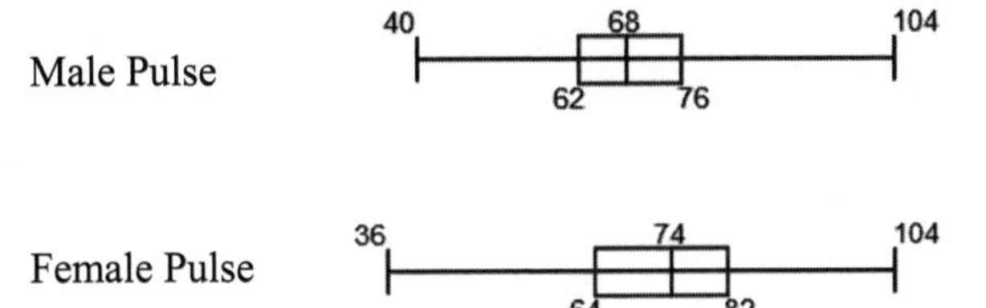

35. The top boxplot represents BMI values for males. The two boxplots do not appear to be very different, so BMI values of males and females appear to be about the same, except for a few very high BMI values for females that caused the boxplot to extend farther to the right.

Male BMI

Female BMI

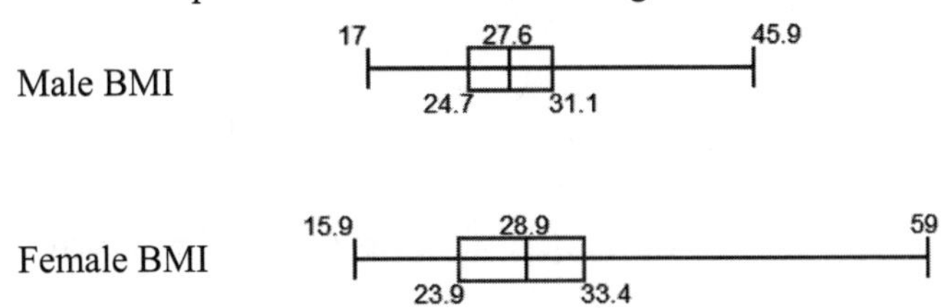

37. Top boxplot represents males. Males appear to have slightly lower pulse rates than females. The outliers for males are 40 beats per minute, 102 beats per minute, and 104 beats per minute. The outlier for females is 36 beats per minute.

Males

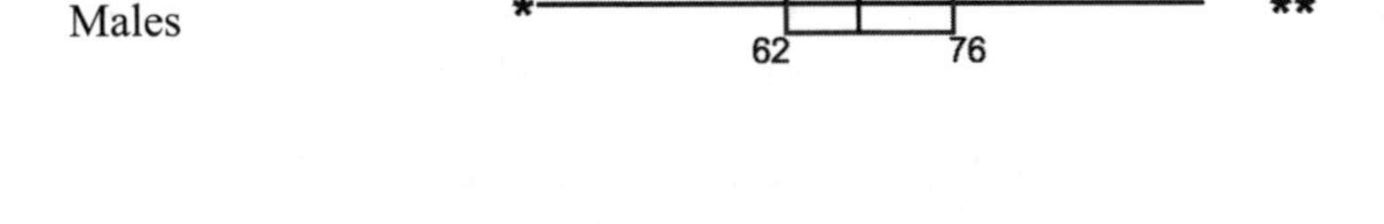

Females

64
82

Quick Quiz

1. The sample mean is $\bar{x} = \frac{4+5+6+6+7+7+7+7+8+8+8+8}{12} = 6.8$ hours.

2. The median is $\frac{7+7}{2} = 7.0$ hours.

3. The modes are 7 hours and 8 hours.

4. The variance is $(1.3 \text{ hour})^2 = 1.7 \text{ hour}^2$

5. Yes, because 0 hours is substantially less than all of the other data values.

6. $z = \dfrac{5-6.3}{1.4} = -0.93$
7. 75% or 60 sleep times
8. minimum, first quartile Q_1, second quartile Q_2 (or median), third quartile Q_3, maximum
9. $s \approx \dfrac{10-4}{4} = 1.5$
10. $\bar{x}, \mu, s, \sigma, s^2, \sigma^2$

Review Exercises

1. a. The mean is $\bar{x} = \dfrac{(-7)+(-7)+(-5)+(-4)+(-1)+0+1+1+1+4+7+22}{12} = 1.0$ min.

   b. The median is $\dfrac{0+1}{2} = 0.5$ min.

   c. The mode is 1 min.

   d. The midrange is $\dfrac{(-7)+22}{2} = 7.5$ min.

   e. The range is $22-(-7) = 29.0$ min.

   f. $s = \sqrt{\dfrac{(-7-1.0)^2+(-7-1.0)^2+(-5-1.0)^2+\cdots+(4-1.0)^2+(7-1.0)^2+(7-1.0)^2}{12-1}} = 7.9$ min

   g. $s^2 = 7.9^2 = 61.8\text{ min}^2$

   h. $L = \dfrac{25\cdot 12}{100} = 3$, so $Q_1 = \dfrac{(-5)+(-4)}{2} = -4.5$ min

   i. $L = \dfrac{75\cdot 12}{100} = 9$, so $Q_1 = \dfrac{1+4}{2} = 2.5$ min (Tech: Minitab yields $Q_1 = -4.75$ min and $Q_3 = 3.25$ min. Excel yields $Q_1 = -4.25$ min and $Q_3 = 1.75$ min.)

2. $z = \dfrac{0-1}{7.9} = -0.13$; The prediction error of 0 minutes is not significant because its z score is between –2 and 2, so it is within two standard deviations of the mean.
3. The five number summary is –7 min, –4.5 min, 1.5 min, 2.5 min, 22 min. (Tech: Minitab yields $Q_1 = -4.75$ min and $Q_3 = 3.25$ min. Excel yields $Q_1 = -4.25$ min and $Q_3 = 1.75$ min.)

-7 0.5 22
-4.5 2.5

4. The mean is $\bar{x} = \dfrac{12+14+22+27+40}{5} = 23.0$. The numbers don't measure or count anything. They are used as replacements for the names of the categories, so the numbers are at the nominal level of measurement. In this case the mean is a meaningless statistic.
5. The male z score is $z = \dfrac{3400-3272.8}{660.2} = 0.19$. The female z score is $z = \dfrac{3200-3037.1}{706.3} = 0.23$. The female has the larger relative birth because the female has the larger z score.
6. The outlier is 646. The mean and standard deviation with the outlier included are $\bar{x} = 267.8$ and $s = 131.6$. With the outlier excluded, the values are $\bar{x} = 230.0$ and $s = 42.0$. Both statistics changed by a substantial amount, so here the outlier has a very strong effect on the mean and standard deviation.
7. The minimum value is 119 mm, the first quartile is 128 mm, the second quartile (or median) is 131 mm, the third quartile is 135 mm, and the maximum value is 141 mm.

8. Based on a minimum of 117 seconds and a maximum of 256 seconds, an estimate of the standard deviation of duration times would be $s \approx \dfrac{256-117}{4} = 34.5$ seconds.

Cumulative Review Exercises

1.

Arsenic (μg)	Frequency
0.0 – 1.9	1
2.0 – 3.9	0
4.0 – 5.9	3
6.0 – 7.9	7
8.0 – 9.9	1

2.

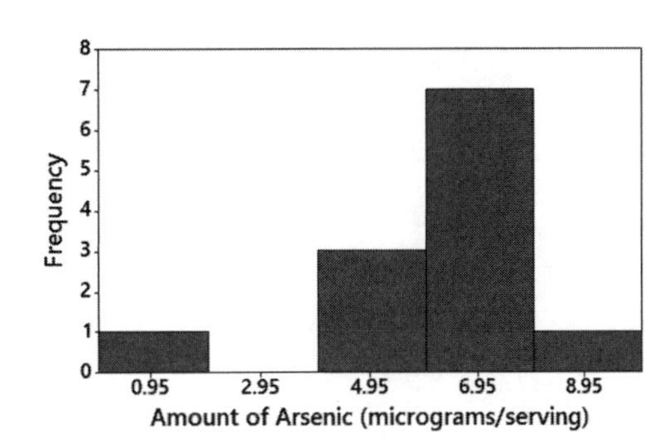

3.

```
1 | 5
2 |
3 |
4 | 9
5 | 44
6 | 13679
7 | 38
8 | 2
```

4. a. The mean is $\bar{x} = \dfrac{1.5+4.9+5.4+5.4+6.1+6.3+6.6+6.7+6.9+7.3+7.8+8.2}{12} = 6.09\ \mu g$.

   b. The median is $\dfrac{6.3+6.6}{2} = 6.45\ \mu g$.

   c. $s = \sqrt{\dfrac{(1.5-6.09)^2+(4.9-6.09)^2+\cdots+(7.8-6.09)^2+(8.2-6.09)^2}{12-1}} = 1.75\ \mu g$

   d. $s^2 = 1.75^2 = 3.06\ (\mu g)^2$

   e. The range is $8.2-1.5 = 6.70\ \mu g$.

5. The vertical scale does not begin at 0, so the differences among different outcomes are exaggerated.
6. No. A normal distribution would appear in a histogram as being bell-shaped, but the histogram is not bell-shaped.

Chapter 4: Probability

Section 4-1: Basic Concepts of Probability

1. $P(A)=\frac{1}{1000}$, or 0.001, $P(\bar{A})=1-\frac{1}{1000}=\frac{999}{1000}$, or 0.999

3. part (c)

5. $0, 3/5, 1, 0.135$

7. $1/9$, or 0.111

9. 47 girls is significantly high.

11. 23 girls is neither significantly low nor significantly high.

13. $1/2$, or 0.5

15. 0.43

17. $1/10$, or 0.1

19. 0

21. $\frac{5}{555}=\frac{1}{111}$, or 0.00901; The employer would suffer because it would be at a risk by hiring someone who uses drugs.

23. $\frac{50}{555}=\frac{10}{111,}$ or 0.0901; This result does appear to be a reasonable estimate of the prevalence rate.

25. $\frac{879}{945}$, or 0.93; Yes, the technique appears to be effective.

27. $\frac{428}{580}=0.738$; Yes, it is reasonable.

29. $\frac{1380}{1380+3732}=\frac{115}{426}$, or 0.270; No, it is not unlikely for someone to not use social networking sites.

31. a. brown /brown, brown/blue, blue/brown, blue/blue
    b. $1/4$
    c. $3/4$

33. $3/8$, or 0.375

35. {bbbb, bbbg, bbgb, bbgg, bgbb, bgbg, bggb, bggg, gbbg, gbbb, gbgb, gbgg, ggbb, ggbg, gggb, gggg}; $4/16$, or 0.25

37. The high probability of 0.327 shows that the sample results could have easily occurred by chance. It appears that there is not sufficient evidence to conclude that pregnant women can correctly predict the gender of their baby.

39. a. In comparing the 200 mg treatment group to the placebo group, the low probability of less than 0.049 shows that the sample results could not have easily occurred by chance. It appears that 200 mg of caffeine does have an effect on memory.
    b. In comparing the 300 mg group to the 200 mg group, the high probability of 0.75 indicates the sample results could have easily occurred by chance. There is not sufficient evidence to conclude that there are different effects from the 300 mg treatment group and the 200 mg treatment group.

41. a. $9999:1$
    b. $4999:1$
    c. The description is not accurate. The odds against winning are $9999:1$ and the odds in favor are $1:9999$, not $1:10,000$.

43. a. $7-2=\$5$
    b. $5:2$
    c. $772:228$ or $193:57$ or about $3.39:1$ (roughly $17:5$)
    d. The worth of the \$2 bet would be approximately $3.39\cdot 2+2=\$8.80$ (instead of the actual payoff of \$7.00).

Section 4-2: Addition Rule and Multiplication Rule

1. $P(A)$ represents the probability of selecting an adult with blue eyes, and $P(\overline{A})$ represents the probability of selecting an adult who does not have blue eyes.
3. Because the selections are made without replacement, the events are dependent. Because the sample size of 1068 is less than 5% of the population size of 15,524,971, the selections can be treated as being independent (based on the 5% guideline for cumbersome calculations).
5. $1-0.26=0.74$
7. $P(\overline{N})=1-0.330=0.670$, where $P(\overline{N})$ is the probability of randomly selecting someone with a response different from "never."
9. $\frac{1118-362}{1118}=\frac{756}{1118}=\frac{387}{559}$, or 0.676
11. $\frac{362}{1118}+\frac{987}{1118}-\frac{329}{1118}=\frac{1020}{1118}=\frac{510}{559}$, or 0.912; The two events are not disjoint.
13. a. $\frac{158}{1118}\cdot\frac{158}{1118}=0.0200$; Yes, the events are independent.
    b. $\frac{158}{1118}\cdot\frac{157}{1117}=0.0199$; The events are dependent, not independent.
15. a. $\frac{987}{1118}\cdot\frac{987}{1118}=0.779$; Yes, the events are independent.
    b. $\frac{987}{1118}\cdot\frac{986}{1117}=0.779$; The events are dependent, not independent.
17. $\frac{362+280}{1118}+\frac{131}{1118}-\frac{33+31}{1118}=\frac{709}{1118}$, or 0.634
19. $\frac{280}{1118}\cdot\frac{279}{1117}\cdot\frac{278}{1116}=0.0156$
21. Use the following table for parts (a) and (b).

	Positive Test Result	Negative Test Result	**Total**
Subject Used Marijuana	True Positive 119	False Negative 3	**122**
Subject Did Not Use Marijuana	False Positive 24	True Negative 154	**178**
Total	**143**	**157**	**300**

    a. There were a total of 300 subjects in the study.
    b. 154 subjects had a true negative result.
    c. $\frac{154}{300}=\frac{77}{150}=0.513$
23. $\frac{119+24+154}{300}=\frac{99}{100}=0.990$
25. a. 0.03
    b. $0.03\cdot 0.03=0.0009$
    c. $0.03\cdot 0.03\cdot 0.03=0.000027$
    d. By using one drive without a backup, the probability of total failure is 0.03, and with three independent disk drives, the probability drops to 0.000027. By changing from one drive to three, the probability of total failure drops from 0.03 to 0.000027, and that is a very substantial improvement in reliability. Back up your data!

27. $8834-504=8330$, $\dfrac{8330}{8834}\cdot\dfrac{8329}{8833}\cdot\dfrac{8328}{8832}=0.838$; The probability of 0.838 is high, so it is likely that the entire batch will be accepted, even though it includes many firmware defects.

29. a. $\dfrac{47{,}637}{47{,}637+111{,}874}=\dfrac{47{,}637}{159{,}511}=0.299$

b. Using the 5% guideline for cumbersome calculations, $(0.299)^5=0.00239$. Using exact probabilities, $\dfrac{47{,}637}{159{,}511}\cdot\dfrac{47{,}636}{159{,}510}\cdot\dfrac{47{,}635}{159{,}509}\cdot\dfrac{47{,}634}{159{,}508}\cdot\dfrac{47{,}633}{159{,}507}=0.00238$.

31. a. $0.985\cdot 0.985+0.985\cdot 0.015+0.015\cdot 0.985=0.999775$

b. $0.985\cdot 0.985=0.970225$

c. The series arrangement provides better protection.

33. $P(A \text{ or } B)=P(A)+P(B)-2P(A \text{ and } B)$

Section 4-3: Complements, Conditional Probability, and Bayes' Theorem

1. $\overline{A}$ is the event of not getting at least 1 defect among the 3 iPhones, which means that all 3 iPhones are good.

3. The probability that the selected person is a high school classmate given that the selected person is female.

5. $1-\left(\dfrac{1}{2}\right)^3=\dfrac{7}{8}$, or 0.875

7. $1-(0.512)^6=0.982$

9. $1-(1-0.10)^4=0.344$

11. $1-(1-0.20)^{15}=0.965$; The probability is high enough so that she can be reasonably sure of getting a defect for her work.

13. a. $P(\text{spent money} \mid \text{given quarters})=\dfrac{27}{43}$, or 0.628

b. $P(\text{kept money} \mid \text{given quarters})=\dfrac{16}{43}$, or 0.372

c. It appears that when students are given four quarters, they are more likely to spend the money than keep it.

15. a. $P(\text{spent money} \mid \text{given quarters})=\dfrac{27}{43}$, or 0.628

b. $P(\text{spent money} \mid \text{given dollar bill})=\dfrac{12}{46}=\dfrac{6}{23}$, or 0.261

c. It appears that students are more likely to spend the money when given four quarters than when given a $1 bill.

17. $P(\text{positive result} \mid \text{no hepatitis C})=\dfrac{2}{1155}$, or 0.00173; This is the probability of the test making it appear that the subject has hepatitis C when the subject does not have it, so the subject is likely to experience needless stress and additional testing.

19. $P(\text{hepatitis C} \mid \text{positive result})=\dfrac{335}{337}$, or 0.994; The very high result makes the test appear to be effective in identifying hepatitis C.

21. a. $1-(0.03)^2=0.9991$

b. $1-(0.03)^3=0.999973$; The usual round-off rule for probabilities would result in a probability of 1.00, which would incorrectly indicate that we are certain to have at least one working hard drive.

23. $1-(1-0.126)^5=0.490$; The probability is not low, so further testing of the individual samples will be necessary for about 49% of the combined samples.

25. $1-\frac{365}{365}\cdot\frac{364}{365}\cdot\frac{363}{365}\cdot\ldots\cdot\frac{341}{365}=1-0.431=0.569$

Section 4-4: Counting

1. The symbol ! is the factorial symbol that represents the product of decreasing whole numbers, as in $6!=6\cdot5\cdot4\cdot3\cdot2\cdot1=720$. Six people can stand in line 720 different ways.

3. Because repetition is allowed, numbers are selected *with replacement*, so the combinations rule and the two permutation rules do not apply. The multiplication counting rule can be used to show that the number of possible outcomes is $10\cdot10\cdot10\cdot10=10{,}000$, so the probability of winning is $1/10{,}000$.

5. $\frac{1}{10}\cdot\frac{1}{10}\cdot\frac{1}{10}\cdot\frac{1}{10}=\frac{1}{10{,}000}$

7. There are ${}_{19}C_2=\frac{19!}{(19-2)!2!}=171$ ways to choose a quinela. The probability is $1/171$, or 0.00585.

9. $8!=40{,}320$; The probability is $1/40{,}320$.

11. There are ${}_{50}P_5=\frac{50!}{(50-5)!}=254{,}251{,}200$ possible routes. The probability is $1/254{,}251{,}200$.

13. $\frac{1}{100\cdot100\cdot100\cdot100}=\frac{1}{100{,}000{,}000}$; No, there are too many different possibilities.

15. ${}_{16}C_4=\frac{16!}{(16-4)!4!}=1820$; The probability is $1/1820$, or 0.000549.

17. $\frac{1}{{}_{69}C_5\cdot26}=\frac{1}{292{,}201{,}338}$

19. $10\cdot10\cdot10\cdot10\cdot10=100{,}000$; The probability is $1/100{,}000$, or 0.00001.

21. There are $8\cdot10\cdot10=800$ possible area codes. There are $8\cdot10\cdot10\cdot8\cdot10\cdot10\cdot10\cdot10\cdot10\cdot10=6{,}400{,}000{,}000$ possible area codes. Yes. (With a total population of about 400,000,000, there would be about 16 phone numbers for every adult and child.)

23. a. ${}_{10}P_4=\frac{10!}{(10-4)!}=5040$

    b. ${}_{10}C_4=\frac{10!}{(10-4)!4!}=210$

    c. The probability is $1/210$.

25. $8!=40{,}320$; The probability is $1/40{,}320$.

27. $\frac{16!}{2!2!2!2!2!}=653{,}837{,}184{,}000$

29. $\frac{1}{{}_{75}C_5\cdot15}=\frac{1}{258{,}890{,}850}$; There is a *much* better chance of being struck by lightning.

31. There are $2+2\cdot2+2\cdot2\cdot2+2\cdot2\cdot2\cdot2+2\cdot2\cdot2\cdot2\cdot2=2+4+8+16+32=62$ different possible characters. The alphabet requires 26 characters and there are 10 digits, so the Morse code system is more than adequate.

33. $\frac{4\cdot16+16\cdot4}{52\cdot51}=\frac{128}{2652}=\frac{32}{663}$, or about 0.0483; 4.83%, or about 5%, of hands are blackjack hands.

35. 12 ways: {25p, 1n 20p, 2n 15p, 3n 10p, 4n 5p, 5n, 1d 15p, 1d 1n 10p, 1d 2n 5p, 1d 3n, 2d 5p,2d 1n} (Note: 25p represents 25 pennies, etc.)

37. $26+26\cdot 36+26\cdot 36^2+26\cdot 36^3+26\cdot 36^4+26\cdot 36^5+26\cdot 36^6+26\cdot 36^7=2{,}095{,}681{,}645{,}538$ or about 2 trillion.

Quick Quiz

1. $\frac{4}{5}$, or 0.8

2. $1-0.20=0.80$

3. $\frac{4}{12}=\frac{1}{3}$

4. $0.74^2=0.5476$, or 0.548

5. Answer varies, but the probability should be low, such as 0.001.

Use the following table for Exercises 6–10.

	Developed Flu	Did Not Develop Flu	**Total**
Vaccine Treatment	14	1056	**1070**
Placebo	95	437	**532**
Total	**109**	**1493**	**1602**

6. $\frac{109}{1602}$, or 0.0680

7. $\frac{14+1056+95}{1602}=\frac{1165}{1602}$, or $\frac{109}{1602}+\frac{1070}{1602}-\frac{14}{1602}=\frac{1165}{1602}$, or 0.727

8. $\frac{14}{1602}=\frac{7}{801}=0.00847$

9. $\frac{109}{1602}\cdot\frac{108}{1601}=0.00459$

10. $P(\text{developed flu} \mid \text{given vaccine})=\frac{14}{1070}=\frac{7}{535}$, or 0.0131

Review Exercises

Use the following table for Exercises 1–10.

	Driver Killed	Driver Not Killed	**Total**
Seatbelt Used	3655	7005	**10,660**
Seatbelt Not Used	4402	3040	**7442**
Total	**8057**	**10,045**	**18,102**

1. $\frac{10{,}660}{18{,}102}=\frac{5330}{9051}$, or 0.589

2. $P(\text{not killed} \mid \text{seatbelt used})=\frac{7005}{10{,}660}=\frac{1401}{2132}$, or 0.657

3. $P(\text{killed} \mid \text{seatbelt not used})=\frac{4402}{7442}=\frac{2201}{3721}$, or 0.592

4. $\frac{10{,}660}{18{,}102}+\frac{8057}{18{,}102}-\frac{3655}{18{,}102}=\frac{7531}{9051}$, or 0.832

5. $\frac{7442}{18{,}102}+\frac{10{,}045}{18{,}102}-\frac{3040}{18{,}102}=\frac{14{,}447}{18{,}102}$, or 0.798

6. $\frac{10{,}660}{18{,}102}\cdot\frac{10{,}659}{18{,}101}=0.347$

7. $\frac{8057}{18{,}102}\cdot\frac{8057}{18{,}102}=0.198$

8. A is the event of selecting a driver and getting someone who was not using a seatbelt. $P(\overline{A})=1-0.589=0.411$

9. A is the event of selecting a driver and getting someone who was killed.
   $P(\bar{A})=1-\dfrac{10{,}045}{18{,}102}=\dfrac{8057}{18{,}102}=\dfrac{1151}{2586}=0.445$

10. $\dfrac{10{,}045}{18{,}102}\cdot\dfrac{10{,}044}{18{,}101}\cdot\dfrac{10{,}043}{18{,}100}=0.171$

11. Answer varies, but Forbes reports that about 19% of cars are black, so any estimate between 0.10 and 0.30 would be good.

12. a. $1-0.75=0.25$, or 25%
    b. $0.75\cdot 0.75\cdot 0.75\cdot 0.75=0.316$
    c. No, it is not unlikely because the probability of 0.316 shows that the event occurs quite often.

13. a. 1/365
    b. 31/365
    c. Answers will vary, but it is probably quite small, such as 0.01 or less.
    d. yes

14. $1-\left(1-\dfrac{34}{10{,}000}\right)^{10}=0.0335$; No, it is not likely.

15. a. $\dfrac{1}{_{33}C_5}=\dfrac{1}{237{,}336}$
    b. 1/4
    c. 1/4
    d. $\dfrac{1}{237{,}336}\cdot\dfrac{1}{4}\cdot\dfrac{1}{4}=\dfrac{1}{3{,}797{,}376}$

16. $\dfrac{1}{_{43}C_5}=\dfrac{1}{962{,}598}$

17. a. $1-\dfrac{1}{1000}=\dfrac{999}{1000}$, or 0.999
    b. $1-\left(\dfrac{1}{1000}\right)^2=\dfrac{999{,}999}{1{,}000{,}000}$, or 0.999999

18. $_{19}P_2=342$; The probability is $1/342$.

Cumulative Review Exercises

1. a. The mean is $\bar{x}=\dfrac{0.09+0.11+\cdots+0.15+0.17+\cdots+0.23+0.35}{12}=0.165$ g/dL.
   b. The median is $\dfrac{0.15+0.17}{2}=0.160$ g/dL.
   c. The midrange is $\dfrac{0.09+0.35}{2}=0.220$ g/dL.
   d. The range is $0.35-0.09=0.260$ g/dL.
   e. $s=\sqrt{\dfrac{(0.09-0.165)^2+(0.11-0.165)^2+\ldots+(0.23-0.165)^2+(0.35-0.165)^2}{12-1}}=0.069$ g/dL
   f. $s^2=(0.069)^2=0.005\ (\text{g/dL})^2$

2. a. The five number summary is 0.090 g/dL, 0.120 g/dL, 0.160 g/dL, 0.180 g/dL, 0.350 g/dL. The value of 0.350 g/dL is an outlier.

   b.

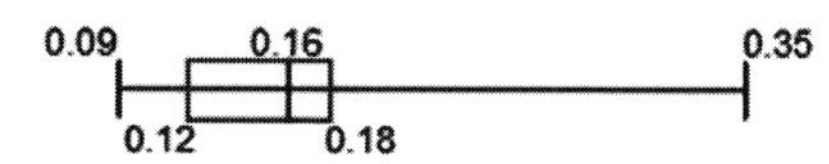

   c.

0. | 9
1. | 113457788
2. | 3
3. | 5

3. a. $\dfrac{2346}{5100} = 0.46 = 46\%$

   b. 0.460

   c. stratified sample

4. a. a convenience sample

   b. If the students at the college are mostly from a surrounding region that includes a large proportion of one ethnic group, the results will not reflect the general population of the United States.

   c. $0.35 + 0.4 = 0.75$

   d. $1 - (0.6)^2 = 0.64$

5. The lack of any pattern of the points in the scatterplot suggests that there does not appear to be an association between systolic blood pressure and blood platelet count.

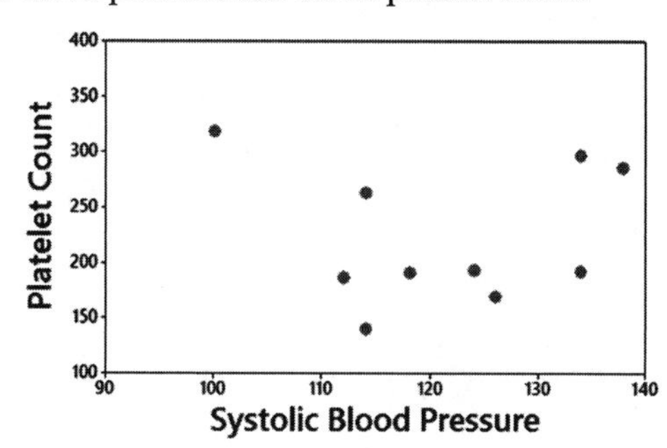

6. a. $\dfrac{1}{_{52}C_5} = \dfrac{1}{2{,}598{,}960}$

   b. $\dfrac{1}{28}$

   c. $\dfrac{1}{2{,}298{,}960} \cdot \dfrac{1}{28} = \dfrac{1}{72{,}770{,}880}$

Chapter 5: Discrete Probability Distributions

Section 5-1: Probability Distributions

1. The random variable is x, which is the number of girls in four births. The possible values of x are 0, 1, 2, 3, and 4. The values of the random variable x are numerical.

3. Table 5-7 does describe a probability distribution because the three requirements are satisfied. First, the variable x is a numerical random variable and its values are associated with probabilities. Second, $\Sigma P(x) = 0.063 + 0.250 + 0.375 + 0.250 + 0.063 = 1.001$, which is not exactly 1 due to round-off error, but is close enough to satisfy the requirement. Third, each of the probabilities is between 0 and 1 inclusive, as required.

5. a. continuous random variable
   b. not a random variable
   c. discrete random variable
   d. continuous random variable
   e. discrete random variable

7. Probability distribution with
   $\mu = 0 \cdot 0.031 + 1 \cdot 156 + 2 \cdot 0.313 + 3 \cdot 313 + 4 \cdot 0.156 + 5 \cdot 0.031 = 2.5$
   $\sigma = \sqrt{(0-2.5)^2 \cdot 0.031 + (1-2.5)^2 \cdot 0.156 + \cdots + (4-2.5)^2 \cdot 0.156 + (5-2.5)^2 \cdot 0.031} = 1.1$

9. Not a probability distribution because the sum of the probabilities is 0.1, which is not 1 as required. Also, Ted clearly needs a new approach.

11. Probability distribution with
    $\mu = 0 \cdot 0.091 + 1 \cdot 0.334 + 2 \cdot 0.408 + 3 \cdot 0.166 = 1.6$
    $\sigma = \sqrt{(0-1.6)^2 \cdot 0.091 + (1-1.6)^2 \cdot 0.334 + (2-1.6)^2 \cdot 0.408 + (3-1.6)^2 \cdot 0.166} = 0.9$
    (The sum of the probabilities is 0.999, but that is due to rounding errors.)

13. This is not a probability distribution because the responses are not values of a numerical random variable.

15. $\mu = 0 \cdot 0.004 + 1 \cdot 0.031 + 2 \cdot 0.109 + \cdots + 6 \cdot 0.109 + 7 \cdot 0.031 + 8 \cdot 0.004 = 4.0$ girls
    $\sigma = \sqrt{(0-4.0)^2 \cdot 0.004 + (1-4.0)^2 \cdot 0.031 + \cdots + (7-4.0)^2 \cdot 0.031 + (8-4.0)^2 \cdot 0.004} = 1.4$ girls

17. The upper limit is $\mu + 2\sigma = 4.0 + 2(1.4) = 6.8$ girls. Because 6 girls is not greater than or equal to 6.8 girls, it is not a significantly high number of girls.

19. a. $P(X = 6) = 0.109$
    b. $P(X \geq 6) = 0.109 + 0.031 + 0.004 = 0.144$
    c. The result from part (b), since it is the probability of the given or more extreme result.
    d. No, because the probability of six or more girls is 0.144, which is not very low (less than or equal to 0.05).

21. $\mu = 0 \cdot 0.172 + 1 \cdot 0.363 + 2 \cdot 0.306 + 3 \cdot 0.129 + 4 \cdot 0.027 + 5 \cdot 0.002 = 1.5$ sleepwalkers
    $\sigma = \sqrt{(0-1.5)^2 \cdot 0.172 + (1-1.5)^2 \cdot 0.363 + \ldots + (4-1.5)^2 \cdot 0.027 + (5-1.5)^2 \cdot 0.002} = 1.0$ sleepwalkers

23. Significantly high numbers of sleepwalkers are greater than or equal to $\mu + 2\sigma = 1.5 + 2(1.0) = 3.5$ sleepwalkers. Because 3 sleepwalkers is not greater than or equal to 3.5 sleepwalkers, 3 sleepwalkers is not a significantly high number.

25. a. $P(X = 1) = 0.363$
    b. $P(X \leq 1) = 0.172 + 0.363 = 0.535$
    c. The probability from part (b), since it is the probability of the given or more extreme result.
    d. No, because the probability of one or fewer sleepwalkers is 0.535, which is not low (less than or equal to 0.05)

27. a. $10 \cdot 10 \cdot 10 = 1000$
    b. $1/1000$
    c. $\$500 - \$1 = \$499$
    d. $-\$1 \cdot 1 + \$500 \cdot (1/1000) = -\$0.50 = -50\,¢$
    e. Because both bets have the same expected value of –50¢, neither bet is better than the other.

29. a. Surviving the year: –\$226; Not surviving the year: \$50,000 – \$226 = \$49,774
    b. $-\$161 \cdot 0.9986 + \$99,839 \cdot (1-0.9986) = -\$21$
    c. Yes; the expected value for the insurance company is \$21, which indicates that the company can expect to make an average of \$21 for each such policy.

Section 5-2: Binomial Probability Distributions

1. The given calculation assumes that the first two consumers are comfortable with the drones and the last three consumers are not comfortable with drones, but there are other arrangements consisting of two consumers who are comfortable and three who are not. The probabilities corresponding to those other arrangements should also be included in the result.
3. Because the 30 selections are made without replacement, they are dependent, not independent. Based on the 5% guideline for cumbersome calculations, the 30 selections can be treated as being independent. (The 30 selections constitute 3% of the population of 1009 responses, and 3% is not more than 5% of the population.) The probability can be found by using the binomial probability formula, but it would require application of that formula 21 times (or 10 times if we are clever), so it would be better to use technology.
5. Not binomial; each of the weights has more than two possible outcomes.
7. binomial
9. Not binomial; because the senators are selected without replacement, the selections are not independent. (The 5% guideline for cumbersome calculations cannot be applied because the 40 selected senators constitute 40% of the population of 100 senators, and that exceeds 5%.)
11. Binomial; although the events are not independent, they can be treated as being independent by applying the 5% guideline. The sample size of 1019 is not more than 5% of the population of all adults.
13. a. $\frac{4}{5} \cdot \frac{4}{5} \cdot \frac{1}{5} = 0.128$
    b. {WWC, WCW, CWW}; Each has a probability of 0.128.
    c. $0.128 \cdot 3 = 0.384$
15. ${}_8C_7 \cdot 0.2^7 \cdot 0.8^1 = 0.0000819$ (Table: 0+)
17. ${}_8C_0 \cdot 0.2^0 \cdot 0.8^8 + {}_8C_1 \cdot 0.2^1 \cdot 0.8^7 + {}_8C_2 \cdot 0.2^2 \cdot 0.8^6 = 0.797$ (Table: 0.798)
19. ${}_8C_0 \cdot 0.2^0 \cdot 0.8^8 = 0.168$
21. ${}_8C_6 \cdot 0.54^6 \cdot 0.46^2 = 0.147$
23. ${}_{10}C_8 \cdot 0.54^8 \cdot 0.46^2 + {}_{10}C_9 \cdot 0.54^9 \cdot 0.46^1 + {}_{10}C_{10} \cdot 0.54^{10} \cdot 0.46^0 = 0.0889$
25. ${}_{90}C_0 \cdot 0.27^0 \cdot 0.73^{90} + {}_{90}C_1 \cdot 0.27^1 \cdot 0.73^{89} + \cdots + {}_{90}C_6 \cdot 0.27^6 \cdot 0.73^{84} + {}_{90}C_7 \cdot 0.27^7 \cdot 0.73^{83} = 0.00000451$; The result of 7 minorities is significantly low. The probability shows that it is very highly unlikely that a process of random selection would result in 7 or fewer minorities. (The Supreme Court rejected the claim that the process was random.)
27. a. ${}_6C_5 \cdot 0.20^5 \cdot 0.80^1 = 0.002$ (Tech: 0.00154)
    b. ${}_6C_6 \cdot 0.20^6 \cdot 0.80^0 = 0+$ (Tech: 0.000064)
    c. $0.002 + 0 = 0.002$ (Tech: 0.00160)
    d. Yes, the small probability from part (c) suggests that 5 is an unusually high number.
29. a. $\mu = np = 36 \cdot 0.5 = 18.0$ girls, $\sigma = \sqrt{np(1-p)} = \sqrt{36 \cdot 0.5 \cdot 0.5} = 3.0$ girls
    b. Values of $18.0 - 2(3.0) = 12.0$ girls or fewer are significantly low, values of $18.0 + 2(3.0) = 24.0$ girls or more are significantly high, and values between 12.0 girls and 24.0 girls are not significant.
    c. The result is significantly high because the result of 26 girls is greater than or equal to 24.0 girls. A result of 26 girls would suggest that the XSORT method is effective.

31. a. $\mu = np = 10 \cdot 0.75 = 7.5$ peas, $\sigma = \sqrt{np(1-p)} = \sqrt{10 \cdot 0.75 \cdot 0.25} = 1.4$ peas

 b. Values of $7.5 - 2(1.4) = 4.7$ peas or fewer are significantly low, values of $7.5 + 2(1.4) = 10.3$ peas or more are significantly high, and values between 4.7 peas and 10.3 peas are not significant.

 c. The result is not significant because the result of 9 peas is not greater than or equal to 10.3 peas.

33. $1 - {}_{36}C_0 \cdot 0.01^0 \cdot 0.99^{36} = 0.304$; It is not unlikely for such a combined sample to test positive.

35. ${}_{40}C_1 \cdot 0.03^1 \cdot 0.97^{39} + {}_{40}C_0 \cdot 0.03^0 \cdot 0.97^{40} = 0.662$; The probability shows that about 2/3 of all shipments will be accepted. With about 1/3 of the shipments rejected, the supplier would be wise to improve quality.

37. a. $\mu = np = 100 \cdot 0.16 = 16.0$ M&Ms, $\sigma = \sqrt{np(1-p)} = \sqrt{100 \cdot 0.16 \cdot 0.84} = 3.7$ M&Ms; Values between $16.0 - 2(3.7) = 8.8$ M&Ms and $16.0 + 2(3.7) = 23.4$ M&Ms are not significant (8.7 and 23.3 if using unrounded values). 19 M&Ms lies between these limits, so it is not significant.

 b. The probability of exactly 19 green M&Ms is ${}_{100}C_{19} \cdot 0.16^{19} \cdot 0.84^{81} = 0.0736$.

 c. The probability of 19 or more green M&Ms is ${}_{100}C_{19} \cdot 0.16^{19} \cdot 0.84^{81} + \cdots + {}_{100}C_{100} \cdot 0.16^{100} \cdot 0.84^{0}$ $= 0.242$.

 d. The probability from part (c) is relevant. The result of 19 green M&Ms is not significantly high.

 e. The results do not provide strong evidence against the claim of 16% for green M&Ms.

39. a. $\mu = np = 611 \cdot 0.43 = 262.7$ votes, $\sigma = \sqrt{np(1-p)} = \sqrt{611 \cdot 0.43 \cdot 0.57} = 12.2$ votes; Values between $262.7 - 2(12.2) = 238.3$ votes and $262.7 + 2(12.2) = 287.1$ votes are not significant. The value of 308 votes is greater than or equal to 128.1, so it is significant.

 b. The probability of exactly 308 voters is ${}_{611}C_{308} \cdot 0.43^{308} \cdot 0.57^{303} = 0.0000369$.

 c. The probability of 308 or more voters is ${}_{611}C_{308} \cdot 0.43^{308} \cdot 0.57^{303} + \cdots + {}_{611}C_{611} \cdot 0.43^{0} \cdot 0.57^{611}$ $= 0.000136$.

 d. The probability from part (c) is relevant. The value of 308 votes is significantly high.

 e. The results suggest that the surveyed voters either lied or had defective memory of how they voted.

41. $P(5) = 0.06(1-0.06)^4 = 0.0468$

43. $P(4) = \dfrac{6!}{(6-2)!2!} \cdot \dfrac{43!}{(43-6+2)!(6-2)!} \div \dfrac{(6+43)!}{(6+43-6)!6!} = 0.1324$

Section 5-3: Poisson Probability Distributions

1. $\mu = 535/576 = 0.929$, which is the mean number of hits per region. $x = 2$, because we want the probability that a randomly selected region had exactly 2 hits, and $e \approx 2.71828$ which is a constant used in all applications of Formula 5-9.

3. The possible values of x are 0, 1, 2, … (with no upper bound), so x is a discrete random variable. It is not possible to have $x = 2.3$ calls in a day

5. a. $P(5) = \dfrac{6.1^5 \cdot e^{-6.1}}{5!} = 0.158$

 b. In 55 years, the expected number of years with 5 hurricanes is $55 \cdot 0.158 = 8.7$.

 c. The expected value of 8.7 years is close to the actual value of 8 years, so the Poisson distribution works well here.

7. a. $P(7) = \dfrac{6.1^7 \cdot e^{-6.1}}{7!} = 0.140$

 b. In 55 years, the expected number of years with 5 hurricanes is $55 \cdot 0.140 = 7.7$.

 c. The expected value of 7.7 years is close to the actual value of 7 years, so the Poisson distribution works well here.

9. $\mu = \dfrac{4221}{365} = 11.6$ births, $P(15) = \dfrac{11.6^{15} \cdot e^{-11.6}}{15!} = 0.0649$ (0.0643 if using the unrounded mean.) There is less than a 7% chance of getting exactly 15 births on any given day.

11. a. $\mu = \dfrac{22713}{365} = 62.2$

 b. $P(50) = \dfrac{62.2^{50} \cdot e^{62.2}}{50!} = 0.0155$

13. a. $P(2) = \dfrac{0.929^2 \cdot e^{-0.929}}{2!} = 0.17$

 b. The expected number of regions with exactly 2 hits is 98.2.

 c. The expected number of regions with 2 hits is close to 93, which is the actual number of regions with 2 hits.

15. $\mu = \dfrac{33{,}561}{2969} = 11.3$ fatalities, $1 - P(0) = 1 - \dfrac{11.3^0 \cdot e^{-11.3}}{0!} = 0.9999876$ or 0.9999877 if using the unrounded mean. There is a very high chance ("almost certain") that at least one fatality will occur.

17. The Poisson distribution approximation is valid since $n = 5200 \geq 100$ and $\mu = np = \dfrac{5200}{292{,}201{,}338}$ $= 0.0000178 \leq 10$. The probability of winning at least one time is $1 - P(0) = 1 - \dfrac{0.0000178^0 \cdot e^{-0.0000178}}{0!}$ $= 0.0000178$, so it is highly unlikely that at least one jackpot win will occur in 50 years.

Quick Quiz

1. No, the sum of the probabilities is $4/3$, or 1.333, which is greater than 1.
2. $\mu = 80 \cdot 0.2 = 16.0;\ \sigma = \sqrt{80 \cdot 0.2 \cdot 0.8} = 3.6$
3. The values are parameters because they represent the mean and standard deviation for the population of all who make random guesses for the 80 questions, not a sample of actual results.
4. No, (Using the range rule of thumb, the limit separating significantly high values is $\mu + 2\sigma = 16.0 + 2(3.6)$ $= 23.2$, but 20 is not greater than or equal to 23.2. Using probabilities, the probability of 20 or more correct answers is 0.163, which is not low.)
5. Yes, (Using the range rule of thumb, the limit separating significantly low values is $\mu - 2\sigma = 16.0 - 2(3.6)$ $= 8.8$, and 8 is less than 8.8. Using probabilities, the probability of 8 or fewer correct answers is 0.0131, which is low.)
6. This is probability distribution because the three requirements are satisfied. First, the variable x is a numerical random variable and its values are associated with probabilities. Second, $\Sigma P(x) =$ $0 + 0.006 + 0.051 + 0.205 + 0.409 + 0.328 = 0.999$, which is not exactly 1 due to round-off error, but is close enough to satisfy the requirement. Third, each of the probabilities is between 0 and 1 inclusive, as required.
7. $\mu = 0 \cdot 0 + 1 \cdot 0.006 + 2 \cdot 0.051 + 3 \cdot 0.205 + 4 \cdot 0.409 + 5 \cdot 0.328 = 4.0$ flights
8. $\sigma^2 = (0.9 \text{ flight})^2 = 0.8 \text{ flight}^2$
9. 0+ indicates that the probability is a very small positive number. It does not indicate that it is impossible for none of the five flights to arrive on time.
10. $P(X < 3) = P(X \leq 2) = 0 + 0.006 + 0.051 = 0.057$

Review Exercises

1. $P(X=3)={}_5C_3 \cdot 0.74^3 \cdot 0.26^2 = 0.247$

2. $P(X \geq 1) = 1 - {}_5C_0 \cdot 0.74^0 \cdot 0.26^5 = 0.999;$ No; the five friends are not randomly selected from the population of adults. Also, the fact that they are vacationing together suggests that their financial situations are more likely to include credit cards.

3. $\mu = 5 \cdot 0.74 = 3.7,\ \sigma = \sqrt{5 \cdot 0.74 \cdot 0.26} = 1.0$.

4. No, the limit separating significantly high values is $\mu + 2\sigma = 3.7 + 2(1.0) = 5.7,$ but 5 is not greater than or equal to 5.7. Also, the probability that all five adults have credit cards is ${}_5C_5 \cdot 0.74^5 \cdot 0.26^0 = 0.222$, which is not low (less than or equal to 0.05).

5. Yes, the limit separating significantly low values is $\mu - 2\sigma = 3.7 - 2(1.0) = 1.7,$ and 1 is less than or equal to 1.7. Also, the probability of one or fewer adults having a credit card is ${}_5C_0 \cdot 0.74^0 \cdot 0.26^5$ $+ {}_5C_1 \cdot 0.74^1 \cdot 0.26^4 = 0.0181$, which is low (less than or equal to 0.05).

6. This is not a probability distribution because the responses are not values of a numerical random variable.

7. This is not a probability distribution because $\Sigma P(x) = 0.0016 + 0.0250 + 0.1432 + 0.3892 + 0.4096$ $= 0.9686,$ instead of 1 as required. The discrepancy between 0.9686 and 1 is too large to attribute to rounding errors.

8. This is a probability distribution (The sum of the probabilities is 0.999, but that is due to rounding errors.) with $\mu = 0 \cdot 0 + 1 \cdot 0.003 + 2 \cdot 0.025 + 3 \cdot 0.111 + 4 \cdot 0.279 + 5 \cdot 0.373 + 6 \cdot 0.208 = 4.6$ people and

   $\sigma = \sqrt{(0-4.6)^2 \cdot 0.0 + (1-4.6)^2 \cdot 0.003 + \cdots + (5-4.6)^2 \cdot 0.373 + (6-4.6)^2 \cdot 0.208} = 1.0$ people.

9. a. $784 \cdot 0.301 = 236.0$ checks

   b. $\mu = 784 \cdot 0.301 = 236.0,\ \sigma = \sqrt{784 \cdot 0.301 \cdot 0.699} = 12.8$

   c. The limit separating significantly low values is $\mu - 2\sigma = 236.0 - 2(12.8) = 210.3$ (210.4 if using unrounded values.)

   d. Yes, because 0 is less than or equal to 210.3 (or 210.4) checks.

10. a. $\mu = 7/365 = 0.0192$

    b. $P(X=0) = \dfrac{0.0192^0 \cdot e^{-0.0192}}{0!} = 0.981$

    c. $P(X>1) = P(X \geq 2) = 1 - P(X \leq 1) = 1 - \dfrac{0.0192^0 \cdot e^{-0.0192}}{0!} - \dfrac{0.0192^1 \cdot e^{-0.0192}}{1!} = 0.000182$

    d. No, because the event is so rare. (But it is possible that more than one death occurs in a car crash or some other such event, so it might be wise to consider a contingency plan.)

Cumulative Review Exercises

1. a. The mean is $\bar{x} = \dfrac{0+0+1+2+8+17+21+28}{8} = 9.6$ moons.

   b. The median is $\dfrac{2+8}{2} = 5.0$ moons.

   c. The mode is 0 moons.

   d. The range is $28 - 0 = 28.0$ moons.

   e. The standard deviation is $s = \sqrt{\dfrac{(0-9.6)^2 + (0-9.6)^2 + \cdots + (21-9.6)^2 + (28-9.6)^2}{8-1}} = 11.0$ moons.

   f. The variance is $s^2 = \dfrac{(0-9.6)^2 + (0-9.6)^2 + \cdots + (21-9.6)^2 + (28-9.6)^2}{8-1} = 120.3$ moons^2.

1. (continued)
   g. The minimum is $9.6-2\cdot 11.0=-12.4$ moons and the maximum is $9.6+2\cdot 11.0=31.6$ moons.
   h. No, because none of the planets have a number of moons less than or equal to $\mu-2\sigma=9.6-2(11.0)=-12.4$ moons (which is impossible, anyway) and none of the planets have a number of moons equal to or greater than $\mu-2\sigma=9.6+2(11.0)=31.6$ moons.
   i. ratio
   j. discrete

2. a. $\frac{1}{10}\cdot\frac{1}{10}\cdot\frac{1}{10}=\frac{1}{1000}=0.001$
   b. $365\cdot 0.001=0.365$
   c. $P(1)=\frac{0.365^1\cdot e^{-0.365}}{1!}=0.254$
   d. $-1\cdot 0.999+499\cdot 0.001=-0.50$ or $-50¢$.

3. Refer to the following table.

	Challenge Upheld with Overturned Call	Challenge Rejected with Overturned Call	**Total**
Challenges by Men	152	412	**564**
Challenges by women	79	236	**315**
Total	**231**	**648**	**879**

   a. $\frac{231}{879}=\frac{77}{293}=0.263$
   b. $\frac{79}{231}=0.342$
   c. $\frac{231}{879}\cdot\frac{230}{878}=0.0688$
   d. $\frac{564}{879}+\frac{231}{879}-\frac{152}{879}=\frac{643}{879}=0.732$
   e. $\frac{152}{231}=0.658$

4. a. $0.73\cdot 347=253$
   b. The sample is the 347 human resource professionals who were surveyed. The population is all human resource professionals.
   c. 73% is a statistic because it is a measure based on a sample, not the entire population.

5. No vertical scale is shown, but a comparison of the numbers shows that 7,066,000 is roughly 1.2 times the number 6,000,000. However, the graph makes it appear that the goal of 7,066,000 people is roughly 3 times the number of people enrolled. The graph is misleading in the sense that it creates the false impression that actual enrollments are far below the goal, which is not the case. Fox News apologized for their graph and provided a corrected graph.

6. a. $P(X=5)={}_8C_5\cdot 0.7^5\cdot 0.3^3=0.254$
   b. $P(X\geq 7)={}_8C_7\cdot 0.7^7\cdot 0.3^1+{}_8C_8\cdot 0.7^8\cdot 0.3^0=0.255$
   c. $\mu=8\cdot 0.7=5.6$ adults, $\sigma=\sqrt{8\cdot 0.7\cdot 0.3}=1.3$ adults
   d. Yes. (Using the range rule of thumb, the limit separating significantly low values is $\mu-2\sigma=5.6-2(1.3)=3$, and 1 is less than 3. Using probabilities, the probability of 1 or fewer people washing their hands is ${}_8C_0\cdot 0.7^0\cdot 0.3^8+{}_8C_1\cdot 0.7^1\cdot 0.3^7=0.00129$, (0.001 if using the table) which is low, such as less than 0.05.

Chapter 6: Normal Probability Distributions

Section 6-1: The Standard Normal Distribution

1. The word "normal" has a special meaning in statistics. It refers to a specific bell-shaped distribution that can be described by Formula 6-1. The lottery digits do not have a normal distribution.
3. The mean and standard deviation have values of $\mu = 0$ and $\sigma = 1$, respectively.
5. $P(x > 3) = 0.2(5-3) = 0.4$
7. $P(2 < x < 3) = 0.2(3-2) = 0.2$
9. $P(z < 0.44) = 0.6700$
11. $P(-0.84 < z < 1.28) = P(z < 1.28) - P(z < -0.84) = 0.8997 - 0.2005 = 0.6992$ (Tech: 0.6993)
13. $z = 1.23$
15. $z = -1.45$
17. $P(z < -1.23) = 0.1093$
19. $P(z < 1.28) = 0.8997$
21. $P(z > 0.25) = 1 - P(z < 0.25) = 1 - 0.5987 = 0.4013$
23. $P(z > -2.00) = 1 - P(z < -2.00) = 1 - 0.0228 = 0.9772$
25. $P(2.00 < z < 3.00) = P(z < 3.00) - P(z < 2.00) = 0.9986 - 0.9772 = 0.0214$ (Tech: 0.0215)
27. $P(-2.55 < z < -2.00) = P(z < -2.00) - P(z < -2.55) = 0.0228 - 0.0054 = 0.0174$
29. $P(-2.00 < z < 2.00) = P(z < 2.00) - P(z < -2.00) = 0.0228 - 0.9772 = 0.9544$ (Tech: 0.9545)
31. $P(-1.00 < z < 5.00) = P(z < 5.00) - P(z < -1.00) = 0.9999 - 0.1587 = 0.8412$ (Tech: 0.8413)
33. $P(z < 4.55) = 0.9999$ (Tech: 0.999997)
35. $P(z > 0) = 0.5000$
37. $P_{99} = 2.33$
39. $P_{2.0} = -2.05$ and $P_{98.0} = 2.05$
41. $z_{0.10} = 1.28$
43. $z_{0.04} = 1.75$
45. $P(-1 < z < 1) = P(z < 1) - P(z < -1) = 0.8413 - 0.1587 = 0.6826 = 68.26\%$ (Tech: 68.27%)
47. $P(-3 < z < 3) = P(z < 3) - P(z < -3) = 0.9987 - 0.0013 = 0.9974 = 99.74\%$ (Tech: 99.73%)

49.a. $P(z > 2) = 1 - P(z < 2) = 1 - 0.9772 = 0.0228 = 2.28\%$

b. $P(z < -2) = 0.0228 = 2.28\%$

c. $P(-2 < z < 2) = P(z < 2) - P(z < -2) = 0.9772 - 0.0228 = 0.9544 = 95.44\%$ (Tech: 95.45%)

Section 6-2: Real Applications of Normal Distributions

1. a. $\mu = 0$ and $\sigma = 1$

   b. The z scores are numbers without units of measurements.
3. The standard normal distribution has a mean of 0 and a standard deviation of 1, but a nonstandard normal distribution has a different value for one or both of those parameters.
5. $z_{x=118} = \dfrac{118-100}{15} = 1.2$; which has an area of 0.8849 to the left.
7. $z_{x=133} = \dfrac{133-100}{15} = 2.2$; which has an area of 0.9861 to the left. $z_{x=79} = \dfrac{110-100}{15} = -1.4$; which has an area of 0.0808 to the left. The area between the two scores is $0.9861 - 0.0808 = 0.9053$.
9. $z = 2.44$; so $x = 2.44 \cdot 15 + 100 = 136$
11. $z = -2.07$; so $x = -2.07 \cdot 15 + 100 = 69$
13. $z_{x=21.0} = \dfrac{21.0-23.5}{1.1} = -2.27$; which has an area of 0.0115 to the left (Table: 0.0116).

15. $z_{x=22.0} = \dfrac{22.0-22.7}{1.0} = -0.70$; which has an area of 0.2420 to the left and $z_{x=24.0} = \dfrac{24.0-22.7}{1.0} = 1.30$; which has an area of 0.9032 to the left. The area between the two scores is $0.9032-0.2420 = 0.6612$.

17. $P_{90} = 1.28$; so the length is $1.28 \cdot 1.1 + 23.5 = 24.9$ in.

19. $P_1 = -2.33$; so the lower length is $-2.33 \cdot 1.1 + 23.5 = 20.9$ in. and $P_{99} = 2.33$; so the upper length is $2.33 \cdot 1.1 + 23.5 = 26.1$ in., so a length of 26 in. is not significantly high.

21. a. $z_{x=78} = \dfrac{78-63.7}{2.9} = 4.93$; which has an area of 0.9999 to the left. $z_{x=62} = \dfrac{62-63.7}{2.9} = -0.59$; which has an area of 0.2776 to the left. Therefore, the percentage of qualified women is $0.9999-0.2776 = 0.7223$, or 72.23% (Tech 72.11%). Yes, about 28% of women are not qualified because of their heights.
    b. The z score with 3% to the left is –1.88, which corresponds to a height of $-1.88 \cdot 2.9 + 63.7 = 58.2$ in. The z score with 3% to the right is 1.88 which corresponds to a height of $1.88 \cdot 2.9 + 63.7 = 69.2$ in.

23. a. The z score for men for the minimum height is $\dfrac{56-68.6}{2.8} = -4.5$ and the z score for men for the maximum height is $\dfrac{56-69.5}{2.4} = -2.36$. The area between the z scores is $0.0091-0.0001 = 0.0090$ or 0.90% (Tech: 0.92%). Because so few men can meet the height requirement, it is likely that most Mickey Mouse characters are women.
    b. The z score with 5% of men to the left is –1.65, which corresponds to a height of $-1.65 \cdot 2.8 + 68.6 = 64.0$ in. and the z score with 50% of men to the right is 0 which corresponds to a height of $0 \cdot 2.8 + 68.6 = 68.6$ in.

25. The z score for eye contact of 230.0 seconds is $\dfrac{230.0-184.0}{55.0} = 0.84$; so 0.2005, or 20.05% (Tech: 20.15%) of people would make eye contact longer than 230 seconds. No, the proportion of schizophrenics is not at all likely to be as high as 0.2005, or about 20%.

27. The z score for the minimum weight is $\dfrac{140.0-171.1}{46.1} = -0.67$ and the z score for the maximum weight is $\dfrac{211.0-171.1}{46.1} = 0.87$. The area between the z scores is $0.8078-0.2514 = 0.5564$, or 55.64% (Tech: 55.67%). Yes, about 44% of women were excluded.

29. a. The z score for the minimum weight is $\dfrac{2495.0-3152.0}{693.4} = -0.95$; which has an area to the left of 0.1711 (Tech: 0.1717).
    b. The z score for 5% to the left is –1.645, which corresponds to a weight of $-1.645 \cdot 693.4 + 3152.0 = 2011.4$ g (Tech: 2011.5 g).
    c. Birth weights are significantly low if they are 2011.4 g or less, and they are "low birth weights" if they are 2495 g or less. Birth weights between 2011.4 g and 2495 g are "low birth weights" but they are not significantly low.

31. a. The z score for a 308 day pregnancy is $\dfrac{308-268}{15} = 2.67$; which corresponds to a probability of 0.0038 or 0.38%. Either a very rare event occurred or the husband is not the father.
    b. The z score corresponding to 3% is –1.87 which corresponds to a pregnancy of $-1.87 \cdot 15 + 268 = 240$ days.

33. a. The mean is 69.5817 (69.6 rounded) beats per minute and the standard deviation is 11.3315 (11.3 rounded) beats per minute. The histogram for the data confirms that the distribution is roughly normal.

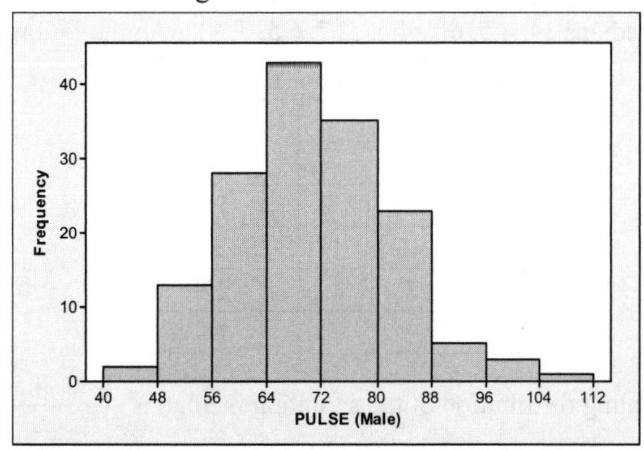

b. The z score for the bottom 2.5% is –1.95, which corresponds to a pulse of $-1.95 \cdot 11.3315 + 69.5817 = 47.4$ beats per minute. The z score for the top 2.5% is 1.95, which corresponds to a pulse of $1.95 \cdot 11.3315 + 69.5817 = 91.8$ beats per minute.

35. a. The new mean is equal to the old mean plus the additional points which is $60 + 15 = 75$. The standard deviation is unchanged at 12 (since the same amount was added to each score.)

b. No, the conversion should also account for variation.

c. The z score for the bottom 70% is 0.52 which has a corresponding score of $60 + 0.52 \cdot 12 = 66.2$, (Tech: 66.3) and the z score for the top 10% is 1.28 which has a corresponding score of $60 + 1.28 \cdot 12 = 75.4$.

d. Using a scheme like the one in part (c), because variation is included in the curving process.

Section 6-3: Sampling Distributions and Estimators

1. a. In the long run, the sample proportions will have a mean of 0.512.

b. The sample proportions will tend to have a distribution that is approximately normal.

3. sample mean, sample variance, sample proportion

5. No, the sample is not a simple random sample from the population of all births worldwide. The proportion of boys born in China is substantially higher than in other countries.

7. a. The population mean is $\mu = \dfrac{4+5+9}{3} = 6$, and the population variance is

$$\sigma^2 = \frac{(4-6)^2 + (5-6)^2 + (9-6)^2}{3} = 4.7.$$

b. The possible sample of size 2 are {(4, 4), (4, 5), (4, 9), (5, 4), (5, 5), (5, 9), (9, 4), (9, 5), (9, 9)} which have the following variances {0, 0.5, 12.5, 0.5, 0, 8, 12.5, 8, 0} respectively.

Sample Variance	Probability
0.0	3/9
0.5	2/9
8	2/9
12.5	2/9

c. The sample variance's mean is $\dfrac{3 \cdot 0 + 2 \cdot 0.5 + 2 \cdot 8 + 2 \cdot 12.5}{9} = 4.7$.

d. Yes. The mean of the sampling distribution of the sample variances (4.7) is equal to the value of the population variance (4.7) so the sample variances target the value of the population variance.

9. a. The population median is 5.
   b. The possible sample of size 2 are {(4, 4), (4, 5), (4, 9), (5, 4), (5, 5), (5, 9), (9, 4), (9, 5), (9, 9)}, which have the following medians {4, 4.5, 6.5, 4.5, 5, 7, 6.5, 7, 9} with the following associated probabilities.

Sample Median	Probability
4	1/9
4.5	2/9
5	1/9
6.5	2/9
7	2/9
9	1/9

   c. The mean of the sampling distribution of the sampling median is $\frac{4+2\cdot 4.5+5+2\cdot 6.52\cdot 7+9}{9} = 6.0$.
   d. No, the mean of the sampling distribution of the sample medians is 6.0, and it is not equal to the value of the population median of 5.0, so the sample medians do not target the value of the population median.

11. a. The possible samples of size 2 are {(34, 34), (34, 36), (34, 41), (34, 51), (36, 34), (36, 36), (36, 41), (36, 51), (41, 34), (41, 36), (41, 41), (41, 51), (51, 34), (51, 36), (51, 41), (51, 51)}, which have the following ranges and associated probabilities.

Sample Range	Probability
34	1/16
35	2/16
36	1/16
37.5	2/16
38.5	2/16
41	2/16
42.5	2/16
43.5	1/16
46	2/16
51	1/16

   b. The mean of the population is $\frac{34+36+41+51}{4} = 40.5$ and the mean of the sample means is $\frac{34+2\cdot 35+36+2\cdot 37.5+2\cdot 38.5+2\cdot 41+2\cdot 42.5+43.5+2\cdot 46+51}{16} = 40.5$ as well.
   c. The sample means target the population mean. Sample means make good estimators of population means because they target the value of the population mean instead of systematically underestimating or overestimating it.

13. a. The possible samples of size 2 are {(34, 34), (34, 36), (34, 41), (34, 51), (36, 34), (36, 36), (36, 41), (36, 51), (41, 34), (41, 36), (41, 41), (41, 51), (51, 34), (51, 36), (51, 41), (51, 51)}, which have the following ranges and associated probabilities.

Sample Range	Probability
0	4/16
2	2/16
5	2/16
7	2/16
10	2/16
15	2/16
17	2/16

13. (continued)

   b. The range of the population is $51-34=17.0$, but the mean of the sample ranges is $\frac{4\cdot 0+2\cdot 2+2\cdot 5+2\cdot 7+2\cdot 10+2\cdot 15+2\cdot 17}{16}=7.0$. The values are not equal.

   c. The sample ranges do not target the population range of 17, so sample ranges do not make good estimators of the population range.

15. The possible birth samples are {(b, b), (b, g), (g, b), (g, g)}.

Proportion of Girls	Probability
0	0.25
1 / 2	0.5
2/2	0.25

Yes. The proportion of girls in 2 births is 0.5, and the mean of the sample proportions is 0.5. The result suggests that a sample proportion is an unbiased estimator of the population proportion.

17. The possibilities are: both questions incorrect, one question correct (two choices), both questions correct.

   a.

Proportion Correct	Probability
$\frac{0}{2}=0$	$\frac{4}{5}\cdot\frac{4}{5}=\frac{16}{25}$
$\frac{1}{2}=0.5$	$2\cdot\left(\frac{1}{5}\cdot\frac{4}{5}\right)=\frac{8}{25}$
$\frac{2}{2}=1$	$\left(\frac{1}{5}\cdot\frac{1}{5}\right)=\frac{1}{25}$

   b. The mean is $\frac{16\cdot 0+8\cdot 0.5+1\cdot 1}{25}=0.2$.

   c. Yes, the sampling distribution of the sample proportions has a mean of 0.2 and the population proportion is also 0.2 (because there is 1 correct answer among 5 choices). Yes, the mean of the sampling distribution of the sample proportions is always equal to the population proportion.

19. The formula yields $P(0)=\frac{1}{2(2-2\cdot 0)!(2\cdot 0)!}=0.25$, $P(0.5)=\frac{1}{2(2-2\cdot 0.5)!(2\cdot 0.5)!}=0.5$, and $P(1)=\frac{1}{2(2-2\cdot 1)!(2\cdot 1)!}=0.25$, which describes the sampling distribution of the sample proportions. The formula is just a different way of presenting the same information in the table that describes the sampling distribution.

Section 6-4: The Central Limit Theorem

1. Because the sample size, n, is greater than 30, the sampling distribution of the mean ages can be approximated by a normal distribution with mean μ and standard deviation $\sigma/\sqrt{40}$.

3. $\mu_{\overline{x}}$ represents the mean of all sample means and $\sigma_{\overline{x}}$ which represents the standard deviation of all sample means. For samples of 64 IQ scores, $\mu_{\overline{x}}=100$, and $\sigma_{\overline{x}}=15/\sqrt{64}=1.875$.

5. a. $z_{x=80}=\frac{80.0-74.0}{12.5}=0.48$; which has a probability of 0.6844 to the left.

   b. $z_{x=80}=\frac{80.0-74.0}{12.5/\sqrt{16}}=1.92$; which has a probability of 0.9726 to the left (Tech: 0.8889).

   c. Because the original population has a normal distribution, the distribution of sample means is a normal distribution for any sample size.

7. a. $z_{x=72} = \dfrac{72.0-74.0}{12.5} = -0.16$ and $z_{x=76} = \dfrac{76.0-74.0}{12.5} = 0.16$; which have a probability of $0.5636 - 0.4364 = 0.1272$ between them (Tech: 0.1271).

   b. $z_{x=72} = \dfrac{72.0-74.0}{12.5/\sqrt{4}} = -0.32$ and $z_{x=76} = \dfrac{76.0-74.0}{12.5/\sqrt{2}} = 0.32$; which have a probability of $0.6255 - 0.3745 = 0.2510$ between them.

   c. Because the original population has a normal distribution, the distribution of sample means is normal for any sample size.

9. $z_{x=185} = \dfrac{185-189}{39/\sqrt{27}} = -0.53$; which has a probability of $1 - 0.2981 = 0.7019$ (Tech: 0.7030) to the right. The elevator does not appear to be safe because there is about a 70% chance that it will be overloaded whenever it is carrying 27 adult males.

11. a. The z score for the top 2% is 2.05, which corresponds to an IQ score of $2.05 \cdot 15 + 100 = 131$.

    b. $z_{x=131} = \dfrac{131-100}{15/\sqrt{4}} = 4.13$; which has a probability of $1 - 0.9999 = 0.0001$ (Tech: 0.0000179) to the right.

    c. No, it is possible that the 4 subjects have a mean of 132 while some of them have scores below the Mensa requirement of 131.

13. a. The mean weight of passengers is $3500/25 = 140$ lb.

    b. $z_{x=140} = \dfrac{140-189}{39/\sqrt{25}} = -6.28$; which has a probability of $1 - 0.0001 = 0.9999$ (Tech: 0.999999998) to the right.

    c. $z_{x=175} = \dfrac{175-189}{39/\sqrt{20}} = -1.61$; which has a probability of $1 - 0.0537 = 0.9463$ (Tech: 0.9458) to the right.

    d. The new capacity of 20 passengers does not appear to be safe enough because the probability of overloading is too high.

15. a. $z_{x=17} = \dfrac{17.0-14.4}{1.0} = 2.60$; which has a probability of $1 - 0.9953 = 0.0047$ to the right.

    b. $z_{x=17} = \dfrac{17.0-14.4}{1.0/\sqrt{122}} = 28.7$; which has a probability of $1 - 0.9999 = 0.0001$ (Tech: 0.0000) to the right.

    c. The result from part (a) is relevant because the seats are occupied by individuals.

17. a. $z_{x=211} = \dfrac{211-171}{46} = 0.87$ and $z_{x=140} = \dfrac{140-171}{46} = -0.67$; which have a probability of $0.8078 - 0.2514 = 0.5564$ between them (Tech: 0.5575).

    b. $z_{x=211} = \dfrac{211-171}{46/\sqrt{25}} = 4.35$ and $z_{x=140} = \dfrac{140-171}{46/\sqrt{25}} = -3.37$; which have a probability of $0.9999 - 0.0004 = 0.9995$ between them (Tech: 0.9996).

    c. Part (a) because the ejection seats will be occupied by individual women, not groups of women.

19. a. $z_{x=72} = \dfrac{72-68.6}{2.8} = 1.21$; which has a probability of 0.8869 (Tech: 0.8877).

    b. $z_{x=72} = \dfrac{72-68.6}{2.8/\sqrt{100}} = 12.1$; which has a probability of 0.9999. (Tech: 1.0000 when rounded to four decimal places.)

    c. The probability of Part (a) is more relevant because it shows that about 89% of male passengers will not need to bend. The result from part (b) gives us useful information about the comfort and safety of individual male passengers.

19. (continued)

    d. Because men are generally taller than women, a design that accommodates a suitable proportion of men will necessarily accommodate a greater proportion of women.

21. a. Yes, the sampling is without replacement and the sample size of 50 is greater than 5% of the finite population size of 275. $\sigma_{\bar{x}} = \frac{16}{\sqrt{50}}\sqrt{\frac{275-50}{275-1}} = 2.0504584$

    b. $z_{x=105} = \frac{105-95.5}{2.0504584} = 4.63$ and $z_{x=95.5} = \frac{95-95.5}{2.0504584} = -0.24$; which have a probability of $1-0.4053 = 0.5947$ between them (Tech: 0.5963).

Section 6-5: Assessing Normality

1. The histogram should be approximately bell-shaped, and the normal quantile plot should have points that approximate a straight-line pattern.
3. We must verify that the sample is from a population having a normal distribution. We can check for normality using a histogram, identifying the number of outliers, and constructing a normal quantile plot.
5. Normal, the points are reasonably close to a straight-line pattern and there is no other pattern that is not a straight-line pattern.
7. Not normal, the points are not reasonably close to a straight-line pattern and there appears to be a pattern that is not a straight-line pattern.
9. normal

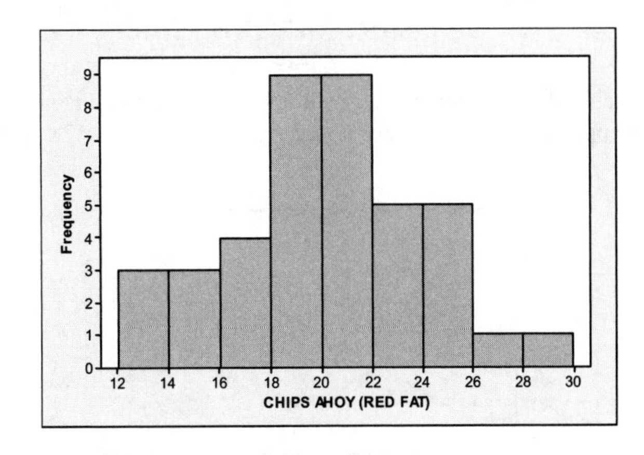

11. not normal

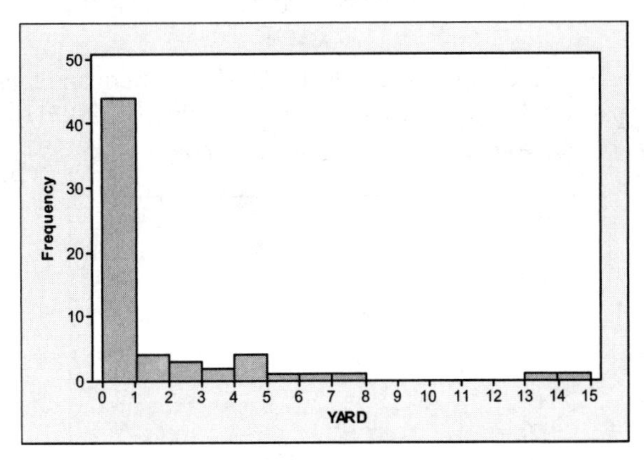

13. Normal, the points are reasonably close to a straight-line pattern and there is no other pattern that is not a straight-line pattern.

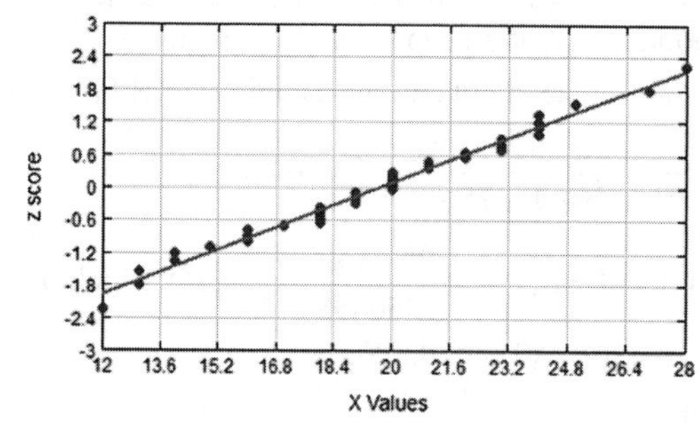

15. Not normal, the points are not reasonably close to a straight-line pattern and there appears to be a pattern that is not a straight-line pattern.

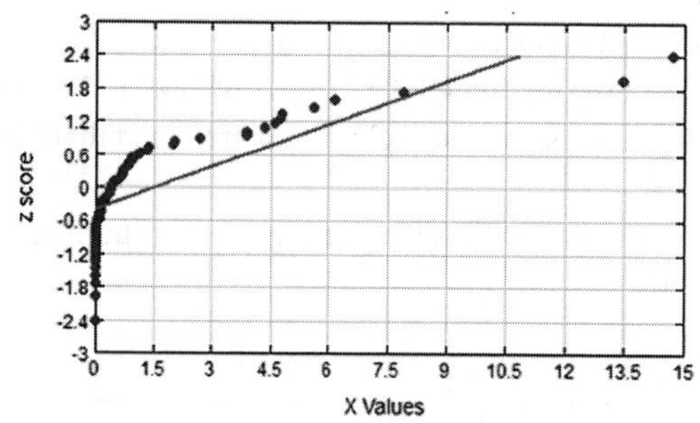

17. Normal, the points are reasonably close to a straight-line pattern and there is no other pattern that is not a straight-line pattern. The points have coordinates $(32.5,-1.28)$, $(34.2,-0.52)$, $(38.5,0)$, $(40.7,0.52)$, and $(44.3,1.28)$.

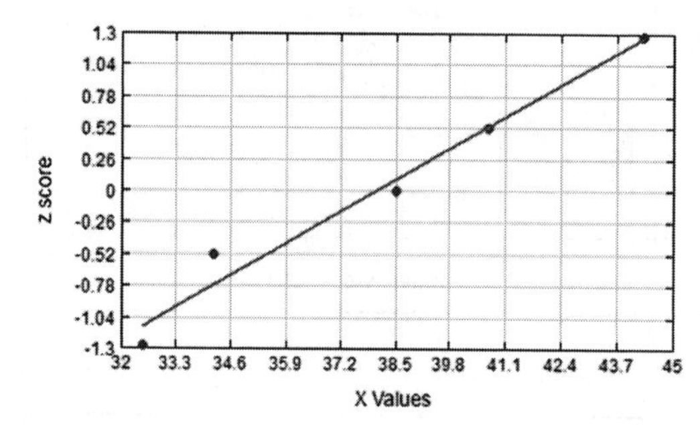

19. Not normal, the points are not reasonably close to a straight-line pattern and there appears to be a pattern that is not a straight-line pattern. The points have coordinates $(963,-1.53)$, $(1027,-0.89)$, $(1029,-0.49)$, $(1034,-0.16)$, $(1070,0.16)$, $(1079,0.49)$, $(1079,0.89)$, and $(1439,1.53)$.

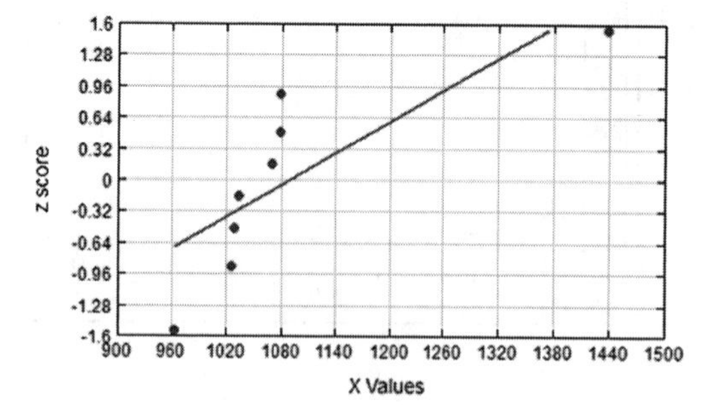

21. a. Yes, the histogram or normal quantile plot will remain unchanged.
    b. Yes, the histogram or normal quantile plot will remain unchanged.
    c. No, the histogram and normal quantile plot will not indicate a normal distribution.

Section 6-6: Normal as Approximation to Binomial

1. a. the area below (to the left of) 502.5
   b. the area between 501.5 and 502.5
   c. the area above (to the right of) 502.5

3. $p = 0.2$, $q = 0.8$, $\mu = 20$, $\sigma = 4$; The value of $\mu = 20$ shows that for people who make random guesses for the 100 questions, the mean number of correct answers is 20. For people who make 100 random guesses, the standard deviation of $\sigma = 4$ is a measure of how much the numbers of correct responses vary.

5. The requirements for the normal approximation are satisfied with $np = 20 \cdot 0.512 = 10.2 \geq 5$ and $nq = 20 \cdot 0.488 = 9.8 \geq 5$. $z_{x=7.5} = \dfrac{7.5 - 20 \cdot 0.512}{\sqrt{20 \cdot 0.512 \cdot 0.488}} = -1.23$; which has a probability of 0.1093 (Tech: 0.1102) to the left.

7. The requirement of $np = 20 \cdot 0.2 = 4 \geq 5$ is not satisfied. The normal approximation should not be used.

9. The requirements for the normal approximation are satisfied with $np = 100 \cdot 0.23 = 23 \geq 5$ and $nq = 100 \cdot 0.77 = 77 \geq 5$. $z_{x=19.5} = \dfrac{19.5 - 100 \cdot 0.23}{\sqrt{100 \cdot 0.23 \cdot 0.77}} = -0.83$; which has a probability of 0.2033 (Tech: 0.2028) to the left. No, 20 is not a significantly low number of white cars. (Tech: Using the binomial distribution: 0.2047.)

11. The requirements for the normal approximation are satisfied with $np = 100 \cdot 0.10 = 10 \geq 5$ and $nq = 100 \cdot 0.90 = 90 \geq 5$. $z_{x=13.5} = \dfrac{13.5 - 100 \cdot 0.10}{\sqrt{100 \cdot 0.10 \cdot 0.90}} = 1.17$ and $z_{x=14.5} = \dfrac{14.5 - 100 \cdot 0.10}{\sqrt{100 \cdot 0.10 \cdot 0.90}} = 1.50$; which have a probability of $0.9524 - 0.9332 = 0.0542$ (Tech: 0.0549) between them. Determination of whether 14 red cars is significantly high should be based on the probability of *14 or more red cars*, not the probability of exactly 14 red cars. (Tech: Using the binomial distribution: 0.0513.)

13. a. The requirements for the normal approximation are satisfied with $np = 879 \cdot 0.25 = 219.75 \geq 5$ and $nq = 879 \cdot 0.75 = 659.25 \geq 5$. $z_{x=230.5} = \dfrac{230.5 - 879 \cdot 0.25}{\sqrt{879 \cdot 0.25 \cdot 0.75}} = 0.84$ and $z_{x=231.5} = \dfrac{231.5 - 879 \cdot 0.25}{\sqrt{879 \cdot 0.25 \cdot 0.75}} = 0.92$; which have a probability of $0.8212 - 0.7995 = 0.0217$ (Tech: 0.0212) between them. (Tech: Using the binomial distribution: 0.0209.)
    b. The requirements for the normal approximation are satisfied with $np = 879 \cdot 0.25 = 219.75 \geq 5$ and $nq = 879 \cdot 0.75 = 659.25 \geq 5$. $z_{x=230.5} = \dfrac{230.5 - 879 \cdot 0.25}{\sqrt{879 \cdot 0.25 \cdot 0.75}} = 0.84$; which has a probability of $1 - 0.7995 = 0.2005$ (Tech: 0.2012) to the right. The result of 231 overturned calls is not significantly high. (Tech: Using the binomial distribution: 0.2006.)

15. a. The requirements for the normal approximation are satisfied with $np = 250 \cdot 0.51 = 127.5 \geq 5$ and $nq = 250 \cdot 0.49 = 122.5 \geq 5$. $z_{x=109.5} = \dfrac{109.5 - 250 \cdot 0.51}{\sqrt{250 \cdot 0.51 \cdot 0.49}} = -2.28$; which has a probability of 0.0113 (Tech: 0.0114) to the left. (Tech: Using the binomial distribution: 0.0113.)
    b. The result of 109 is significantly low.

17. a. The requirements for the normal approximation are satisfied with $np = 929 \cdot 0.75 = 696.75 \geq 5$ and $nq = 929 \cdot 0.25 = 232.25 \geq 5$. $z_{x=704.5} = \dfrac{704.5 - 929 \cdot 0.75}{\sqrt{929 \cdot 0.75 \cdot 0.25}} = 0.59$; which has a probability of $1 - 0.7224 = 0.2776$ (Tech: 0.2785) to the right. (Tech: Using the binomial distribution: 0.2799.)

17. (continued)
    b. The result of 705 peas with red flowers is not significantly high.
    c. The result of 705 peas with red flowers is not strong evidence against Mendel's assumption that $3/4$ of peas will have red flowers.
19. a. Using the normal approximation: $\mu = 1002 \cdot 0.61 = 611.22$, $\sigma = \sqrt{1002 \cdot 0.61 \cdot 0.39} = 15.4394$, and $z_{x=700.5} = \dfrac{700.5 - 611.22}{\sqrt{1002 \cdot 0.61 \cdot 0.39}} = 5.78$; which has a probability of 0.0001 (Tech 0.0000) to the right.
    b. The result suggests that the surveyed people did not respond accurately.
21. (1) The requirements for the normal approximation are satisfied with $np = 11 \cdot 0.512 = 5.632 \geq 5$ and $nq = 11 \cdot 0.488 = 5.368 \geq 5$. $z_{x=6.5} = \dfrac{6.5 - 11 \cdot 0.512}{\sqrt{11 \cdot 0.512 \cdot 0.488}} = 0.52$ and $z_{x=7.5} = \dfrac{7.5 - 11 \cdot 0.512}{\sqrt{11 \cdot 0.512 \cdot 0.488}} = 1.13$; which have a probability of $0.8708 - 0.6985 = 0.1723$ between them.
    (2) 0.1704
    (3) 0.1726
    No, the approximations are not off by very much.

Quick Quiz

1.

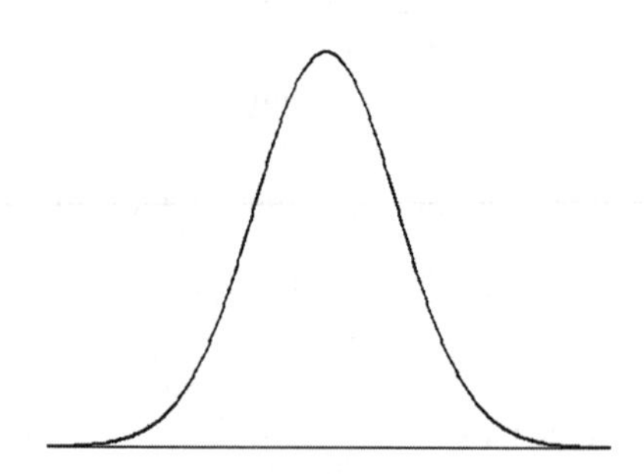

2. $z = P_9 = -1.34$

3. $P(z > -2.93) = 1 - P(z < -2.93) = 1 - 0.0017 = 0.9983$

4. $P(0.87 < z < 1.78) = P(z < 1.78) - P(z < 0.87) = 0.9625 - 0.8078 = 0.1547$ (Tech: 0.1546)

5. a. $\mu = 0$ and $\sigma = 1$
   b. $\mu_{\bar{x}}$ represents the mean of all sample means, and $\sigma_{\bar{x}}$ represents the standard deviation of all sample means.

6. $z_{x=80} = \dfrac{80.0 - 70.2}{11.2} = 0.88$; which has a probability of 0.8106 (Tech: 0.8092) to the left.

7. $z_{x=60} = \dfrac{60.0 - 70.2}{11.2} = -0.91$ and $z_{x=80} = \dfrac{80.0 - 70.2}{11.2} = 0.88$; which have a probability of $0.8106 - 0.1814$ $= 0.6292$ (Tech: 0.6280) between them.

8. The z score for P_{90} is 1.28, which corresponds to a diastolic blood pressure of $1.28 \cdot 11.2 + 70.2 = 84.5$ mm Hg (Tech: 84.6 mm Hg).

9. $z_{x=75} = \dfrac{75.0 - 70.2}{11.2/\sqrt{16}} = 1.71$; which has a probability of 0.9564 (Tech: 0.9568) to the left.

10. The normal quantile plot suggests that diastolic blood pressure levels of women are normally distributed.

Review Exercises

1. a. The probability to the left of a z score of 1.54 is 0.9382.
   b. The probability to the right of a z score of –1.54 is $1-0.0618=0.9382$.
   c. The probability between z scores –1.33 and 2.33 is $0.9901-0.0918=0.8983$.
   d. The z score for Q_1 is –0.67.
   e. $z_{x=0.50}=\dfrac{0.50-0.0}{1/\sqrt{9}}=1.50$; which has a probability of $1-0.9332=0.0668$ to the right.

2. a. $z_{x=54}=\dfrac{54-59.7}{2.5}=-2.28$; which has a probability of 0.0113 to the left, or 1.13%.
   b. The z score for the lowest 95% is 1.645 which corresponds to a standing eye height of $1.645\cdot 2.5+59.7=63.8$ in.

3. a. $z_{x=70}=\dfrac{70-64.3}{2.6}=2.19$; which has a probability of $1-0.9857=0.0143$ to the right, or 1.43%. (Tech: 1.42%)
   b. The z score for the lowest 2% is –2.05 which corresponds to a standing eye height of $-20.05\cdot 2.6+64.3=59.0$ in.

4. a. The distribution of samples means is normal.
   b. $\mu_{\bar{x}}=100$
   c. $\sigma_{\bar{x}}=15/\sqrt{64}=1.875$

5. a. An unbiased estimator is a statistic that targets the value of the population parameter in the sense that the sampling distribution of the statistic has a mean that is equal to the mean of the corresponding parameter.
   b. mean, variance and proportion
   c. true

6. a. $z_{x=72}=\dfrac{72-68.6}{2.8}=1.21$; which has a probability of 0.8869, or 88.69% (Tech: 88.77%) to the left. With about 11% of all men needing to bend, the design does not appear to be adequate, but the Mark VI monorail appears to be working quite well in practice.
   b. The z score for 99% is 2.33 which corresponds to a doorway height of $2.33\cdot 2.8+68.6=75.1$ in.

7. a. Because women are generally a little shorter than men, a doorway height that accommodates men will also accommodate women.
   b. $z_{x=72}=\dfrac{72-68.6}{2.8/\sqrt{60}}=9.41$; which has a probability of 0.9999, or 1 when rounded.
   c. Because the mean height of 60 men is less than 72 in., it does not follow that the 60 individual men all have heights less than 72 in. In determining the suitability of the door height for men, the mean of 60 heights is irrelevant, but the heights of individual men are relevant.

8. a. No, a histogram is far from bell shaped and a normal quantile plot reveals a pattern of points that is far from a straight-line pattern.

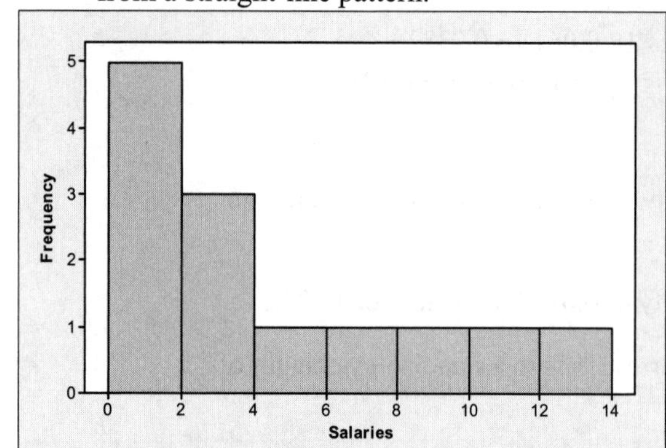

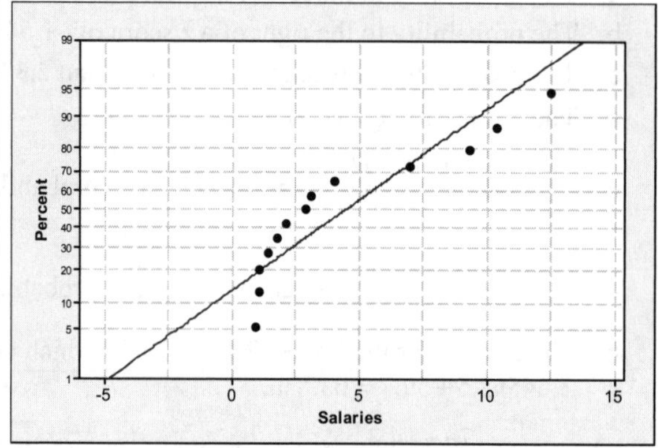

b. No, the sample size $(n = 13)$ does not satisfy the condition of $n > 30$ and the values do not appear to be from a population having a normal distribution.

9. The requirements for the normal approximation are satisfied with $np = 1064 \cdot 0.75 = 798 \geq 5$ and $nq = 1064 \cdot 0.25 = 266 \geq 5$. $z_{z=787.5} = \dfrac{787.5 - 1064 \cdot 0.75}{\sqrt{11064 \cdot 0.75 \cdot 0.25}} = -0.74$; which has a probability of 0.2296 (Tech: 0.2286) to the left. The occurrence of 787 offspring plants with long stem is not unusually low because its probability is not small. The results are consistent with Mendel's claimed proportion of $3/4$. (Tech: Using the binomial distribution: 0.2278.)

10. a. $z_{x=70} = \dfrac{70.0 - 63.7}{2.9} = 2.17$; which has a probability of $1 - 0.9850 = 0.0150$, or 1.50% (Tech: 1.49%) to the right.

b. The z score for the upper 2.5% is 1.96 which corresponds to a doorway height of $1.96 \cdot 2.9 + 63.7 = 69.4$ in.

Cumulative Review Exercises

1. a. The mean is $\bar{x} = \dfrac{0.6 + 0.9 + \cdots + 1.7 + 2.6 + 2.8 + 3.3 + 4.1 + \cdots + 22.5 + 23.4}{15} = \$6.02$ million, or \$6,020,000.

b. The median is \$2.80 million, or \$2,800,000.

c. $s = \sqrt{\dfrac{(0.6 - 6.02)^2 + (0.9 - 6.02)^2 + \ldots + (22.5 - 6.02)^2 + (23.4 - 6.02)^2}{15 - 1}} = \$7.47$ million, or \$7,470,000.

d. $s^2 = (7.47)^2 = 55.73$ (million dollars)^2

e. $z_{x=23.4} = \dfrac{23.4 - 6.02}{7.47} = 2.33$

f. ratio

g. discrete

2. a. Q_1: $L = \dfrac{25 \cdot 15}{100} = 3.75$, so $Q_1 = \$1.30$ million

Q_2: $L = \dfrac{50 \cdot 15}{100} = 7.5$, so $Q_2 = \$2.80$ million

Q_3: $L = \dfrac{75 \cdot 15}{100} = 11.25$, so $Q_2 = \$7.10$ million

2. (continued)

   b.

   c. The sample does not appear to be from a population having a normal distribution.

3. No, the distribution does not appear to be a normal distribution.

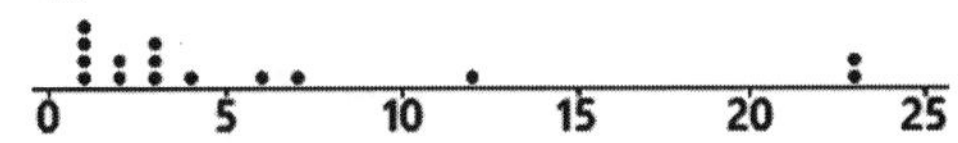

4. a. $\overline{B}$ is the event of selecting someone who does not have blue eyes.

   b. $P(\overline{B}) = 1 - 0.35 = 0.65$

   c. $0.35 \cdot 0.35 \cdot 0.35 = 0.0429$

   d. The requirements for the normal approximation are satisfied with $np = 100 \cdot 0.35 = 35 \geq 5$ and $nq = 100 \cdot 0.65 = 65 \geq 5$. $z_{x=39.5} = \dfrac{39.5 - 100 \cdot 0.35}{\sqrt{100 \cdot 0.35 \cdot 0.65}} = 0.94$; which has a probability of $1 - 0.8264 = 0.1736$ (Tech: 0.1727) to the right. (Tech: Using the binomial distribution: 0.1724.)

   e. No, 40 people with blue eyes is not significantly high.

5. a. $z_{x=10} = \dfrac{10 - 9.6}{0.5} = 0.80$; which has a probability of 0.7881 to the left.

   b. $z_{x=8.0} = \dfrac{8.0 - 9.6}{0.5} = -3.20$ and $z_{x=11.0} = \dfrac{11.0 - 9.6}{0.5} = 2.80$; which have a probability of $0.9974 - 0.0007 = 0.9967$ (Tech: 0.9968) between them.

   c. The z score for the top 5% is 1.645, which correspond to the length $1.645 \cdot 0.5 + 9.6 = 10.4$ in.

   d. $z_{x=9.8} = \dfrac{9.8 - 9.6}{0.5/\sqrt{25}} = 2.00$; which has a probability of $1 - 0.9772 = 0.0228$ to the right.

Chapter 7: Estimating Parameters and Determining Sample Sizes

Section 7-1: Estimating a Population Proportion

1. The confidence level (such as 95%) was not provided.

3. $\hat{p}=0.14$ is the sample proportion; $\hat{q}=0.86$ (found from evaluating $1-\hat{p}$); $n=1000$ is the sample size; $E=0.04$ is the margin of error; p is the population proportion, which is unknown. The value of α is 0.05.

5. $z_{0.05}=1.645$

7. $z_{0.0025}=2.81$

9. $\hat{p}=\dfrac{0.0434+0.217}{2}=0.130$ and $E=\dfrac{0.217-0.0434}{2}=0.087$; 0.130 ± 0.087

11. $0.0169<p<0.143$

13. a. $\hat{p}=33/362=0.0912$

 b. $E=z_{\alpha/2}\sqrt{\dfrac{\hat{p}\hat{q}}{n}}=1.96\sqrt{\dfrac{\left(\frac{33}{362}\right)\left(\frac{329}{362}\right)}{362}}=0.0297$

 c. $\hat{p}-E<p<\hat{p}-E$

 $0.0912-0.0297<p<0.0912+0.0297$

 $0.0615<p<0.121$

 d. We have 95% confidence that the interval from 0.0615 to 0.121 actually does contain the true value of the population proportion of McDonald's drive-through orders that are not accurate.

15. a. $\hat{p}=717/5000=0.143$

 b. $E=z_{\alpha/2}\sqrt{\dfrac{\hat{p}\hat{q}}{n}}=1.645\sqrt{\dfrac{\left(\frac{717}{5000}\right)\left(\frac{4283}{5000}\right)}{5000}}=0.00815$

 c. $\hat{p}-E<p<\hat{p}-E$

 $0.143-0.00815<p<0.143+0.00815$

 $0.135<p<0.152$

 d. We have 90% confidence that the interval from 0.135 to 0.152 actually does contain the true value of the population proportion of returned surveys.

17. 95% CI: $\hat{p}\pm z_{\alpha/2}\sqrt{\dfrac{\hat{p}\hat{q}}{n}}=\dfrac{426}{860}\pm 1.96\sqrt{\dfrac{\left(\frac{426}{860}\right)\left(\frac{434}{860}\right)}{860}}\Rightarrow 0.462<p<0.529$; Because 0.512 is contained within the confidence interval, there is not strong evidence against 0.512 as the value of the proportion of boys in all births.

19. 99% CI: a. $\hat{p}\pm z_{\alpha/2}\sqrt{\dfrac{\hat{p}\hat{q}}{n}}=\dfrac{54}{318}\pm 2.58\sqrt{\dfrac{\left(\frac{54}{318}\right)\left(\frac{264}{318}\right)}{318}}\Rightarrow 0.116<p<0.224$, or $11.6\%<p<22.4\%$

 b. Because the two confidence intervals overlap, it is possible that Burger King and Wendy's have the same rate of orders that are not accurate. Neither restaurant appears to have a significantly better rate of accuracy of orders.

21. a. 0.5

 b. $\hat{p}=123/280=0.439$

 c. 99% CI: $\hat{p}\pm z_{\alpha/2}\sqrt{\dfrac{\hat{p}\hat{q}}{n}}=\dfrac{123}{280}\pm 2.56\sqrt{\dfrac{\left(\frac{123}{280}\right)\left(\frac{157}{280}\right)}{280}}\Rightarrow 0.363<p<0.516$

 d. If the touch therapists really had an ability to select the correct hand by sensing an energy field, their success rate would be significantly greater than 0.5, but the sample success rate of 0.439 and the confidence interval suggest that they do not have the ability to select the correct hand by sensing an energy field.

23. a. 90% CI: $\hat{p} \pm z_{\alpha/2}\sqrt{\dfrac{\hat{p}\hat{q}}{n}} = 0.000321 \pm 1.645\sqrt{\dfrac{(0.000321)(0.999679)}{420{,}095}} \Rightarrow 0.000276 < p < 0.000366$, or $0.0276\% < p < 0.0366\%$; (Using $x = 135$: $0.0276\% < p < 0.0367\%$)

    b. No, because 0.0340% is included in the confidence interval.

25. Placebo group:

    95% CI: $\hat{p} \pm z_{\alpha/2}\sqrt{\dfrac{\hat{p}\hat{q}}{n}} = \dfrac{7}{270} \pm 1.96\sqrt{\dfrac{\left(\frac{7}{270}\right)\left(\frac{263}{270}\right)}{270}} \Rightarrow 0.00697 < p < 0.0449$, or $0.697\% < p < 4.49\%$

    Treatment group:

    95% CI: $\hat{p} \pm z_{\alpha/2}\sqrt{\dfrac{\hat{p}\hat{q}}{n}} = \dfrac{8}{863} \pm 1.96\sqrt{\dfrac{\left(\frac{7}{863}\right)\left(\frac{855}{863}\right)}{863}} \Rightarrow 0.00288 < p < 0.0157$, or $0.288\% < p < 1.57\%$

    Because the two confidence intervals overlap, there does not appear to be a significant difference between the rates of allergic reactions. Allergic reactions do not appear to be a concern for Lipitor users.

27. Sustained care:

    95% CI: $\hat{p} \pm z_{\alpha/2}\sqrt{\dfrac{\hat{p}\hat{q}}{n}} = 0.828 \pm 1.96\sqrt{\dfrac{(0.828)(0.172)}{198}} \Rightarrow 0.775 < p < 0.881$, or $77.5\% < p < 88.1\%$;

    (Using $x = 164$: $77.5\% < p < 88.1\%$)

    Standard care:

    95% CI: $\hat{p} \pm z_{\alpha/2}\sqrt{\dfrac{\hat{p}\hat{q}}{n}} = 0.628 \pm 1.96\sqrt{\dfrac{(0.628)(0.372)}{199}} \Rightarrow 0.561 < p < 0.695$, or $56.1\% < p < 69.5\%$;

    The two confidence intervals do not overlap. It appears that the success rate is higher with sustained care.

29. $\hat{p} = 18/34 = 0.529$, 95% CI: $\hat{p} \pm z_{\alpha/2}\sqrt{\dfrac{\hat{p}\hat{q}}{n}} = \dfrac{18}{34} \pm 1.96\sqrt{\dfrac{\left(\frac{18}{34}\right)\left(\frac{16}{34}\right)}{34}} \Rightarrow 0.362 < p < 0.697$,

    or $36.2\% < p < 69.7\%$; Greater height does not appear to be an advantage for presidential candidates. If greater height is an advantage, then taller candidates should win substantially more than 50% of the elections, but the confidence interval shows that the percentage of elections won by taller candidates is likely to be anywhere between 36.2% and 69.7%.

31. a. $n = \dfrac{\left[z_{\alpha/2}\right]^2 \hat{p}\hat{q}}{E^2} = \dfrac{[2.575]^2 \cdot 0.25}{0.03^2} = 1842$ (Tech: 1844)

    b. $n = \dfrac{\left[z_{\alpha/2}\right]^2 \hat{p}\hat{q}}{E^2} = \dfrac{[2.575]^2 \cdot (0.10)(0.90)}{0.03^2} = 664$

    c. They don't change.

33. a. $n = \dfrac{\left[z_{\alpha/2}\right]^2 \hat{p}\hat{q}}{E^2} = \dfrac{[1.96]^2 \cdot 0.25}{0.05^2} = 385$

    b. $n = \dfrac{\left[z_{\alpha/2}\right]^2 \hat{p}\hat{q}}{E^2} = \dfrac{[1.96]^2 \cdot (0.40)(0.60)}{0.05^2} = 369$

    c. No, the sample size doesn't change much.

35. a. $n = \dfrac{\left[z_{\alpha/2}\right]^2 \hat{p}\hat{q}}{E^2} = \dfrac{[1.645]^2 \cdot 0.25}{0.025^2} = 1083$

    b. $n = \dfrac{\left[z_{\alpha/2}\right]^2 \hat{p}\hat{q}}{E^2} = \dfrac{[1.645]^2 \cdot (0.38)(0.62)}{0.025^2} = 1021$

37. a. $n = \dfrac{[z_{\alpha/2}]^2 \hat{p}\hat{q}}{E^2} = \dfrac{[1.645]^2 \cdot 0.25}{0.03^2} = 752$

b. $n = \dfrac{[z_{\alpha/2}]^2 \hat{p}\hat{q}}{E^2} = \dfrac{[1.645]^2 \cdot (0.16)(0.84)}{0.03^2} = 405$

c. No. A sample of the people you know is a convenience sample, not a simple random sample, so it is very possible that the results would not be representative of the population.

39. $n = \dfrac{N\hat{p}\hat{q}[z_{\alpha/2}]^2}{\hat{p}\hat{q}[z_{\alpha/2}]^2 + (N-1)E^2} = \dfrac{2500(0.82)(0.18)[2.575]^2}{(0.82)(0.18)[2.575]^2 + (2500-1)0.02^2} = 1237$ (Tech: 1238)

41. a. The requirement of at least 5 successes and at least 5 failures is not satisfied, so the normal distribution cannot be used.

b. $3/40 = 0.075$

Section 7-2: Estimating a Population Mean

1. a. 13.05 Mbps $< \mu <$ 22.15 Mbps

b. The best point estimate of μ is $\bar{x} = \dfrac{13.046 + 22.15}{2} = 17.60$ Mbps. The margin of error is $E = \dfrac{22.15 - 13.046}{2} = 4.55$ Mbps.

c Because the sample size of 50 is greater than 30, we can consider the sample mean to be from a population with a normal distribution.

3. We have 95% confidence that the limits of 13.05 Mbps and 22.15 Mbps contain the true value of the mean of the population of all Verizon data speeds at the airports.

5. Neither the normal nor the t distribution applies.

7. $z_{\alpha/2} = 2.576$ (Table: 2.575)

9. The sample size is greater than 30 and the data appear to be from a population that is normally distributed. 95% CI: $\bar{x} \pm t_{\alpha/2}\dfrac{s}{\sqrt{n}} = 30.4 \pm 1.972 \cdot \dfrac{7.1}{\sqrt{205}} \Rightarrow 29.4 \text{ hg} < \mu < 31.4 \text{ hg}$; No, the results do not differ by much.

11. The sample size is greater than 30 and the data appear to be from a population that is normally distributed. 95% CI: $\bar{x} \pm t_{\alpha/2}\dfrac{s}{\sqrt{n}} = 98.20 \pm 1.983 \cdot \dfrac{0.62}{\sqrt{106}} \Rightarrow 98.08°\text{F} < \mu < 98.32°\text{F}$; Because the confidence interval does not contain 98.6°F, it appears that the mean body temperature is not 98.6°F, as is commonly believed.

13. It is assumed that the 16 sample values appear to be from a normally distributed population. 98% CI: $\bar{x} \pm t_{\alpha/2}\dfrac{s}{\sqrt{n}} = 98.9 \pm 2.602 \cdot \dfrac{42.3}{\sqrt{16}} \Rightarrow 71.4 \text{ min} < \mu < 126.4 \text{ min}$; The confidence interval includes the mean of 102.8 min that was measured before the treatment, so the mean could be the same after the treatment. This result suggests that the zopiclone treatment does not have a significant effect.

15. The sample appears to have a normal distribution. 95% CI: $\bar{x} \pm t_{\alpha/2}\dfrac{s}{\sqrt{n}} = 2.6 \pm 2.262 \cdot \dfrac{1.07}{\sqrt{10}}$ $\Rightarrow 1.8 < \mu < 3.4$; The given numbers are just substitutes for the four DNA base names, so the numbers don't measure or count anything, and they are at the nominal level of measurement. The confidence interval has no practical use.

17. The sample appears to have a normal distribution. 99% CI: $\bar{x} \pm t_{\alpha/2}\dfrac{s}{\sqrt{n}} = 7.0 \pm 3.707 \cdot \dfrac{2.216}{\sqrt{7}}$ $\Rightarrow 5.0 < \mu < 9.0$; The results tell us nothing about the population of adult females.

19. The sample appears to have a normal distribution. 98% CI: $\bar{x} \pm t_{\alpha/2}\frac{s}{\sqrt{n}} = 0.719 \pm 3.143 \cdot \frac{0.366}{\sqrt{7}}$ $\Rightarrow 0.284 \text{ ppm} < \mu < 1.153 \text{ ppm}$; Using the FDA guideline, the confidence interval suggests that there could be too much mercury in fish because it is possible that the mean is greater than 1 ppm. Also, one of the sample values exceeds the FDA guideline of 1 ppm, so at least some of the fish have too much mercury.

21. The data appear to have a distribution that is far from normal, so the confidence interval might not be a good estimate of the population mean.
98% CI: $\bar{x} \pm t_{\alpha/2}\frac{s}{\sqrt{n}} = 172 \pm 2.821 \cdot \frac{32.2}{\sqrt{10}} \Rightarrow 143.3 \text{ million dollars} < \mu < 200.7 \text{ million dollars}$; Because the amounts are from the ten wealthiest celebrities, the confidence interval doesn't tell us anything about the population of all celebrities.

23. The sample appears to have a normal distribution. 90% CI: $\bar{x} \pm t_{\alpha/2}\frac{s}{\sqrt{n}} = 3.92 \pm 1.761 \cdot \frac{0.549}{\sqrt{15}}$ $\Rightarrow 3.67 < \mu < 4.17$; Because all of the students were at the University of Texas at Austin, the confidence interval doesn't tell us anything about the population of college students in Texas.

25. Both samples appear to have a normal distribution.
Males: 95% CI: $\bar{x} \pm t_{\alpha/2}\frac{s}{\sqrt{n}} = 69.58 \pm 1.976 \cdot \frac{0.916}{\sqrt{153}} \Rightarrow 67.8 \text{ bpm} < \mu < 71.4 \text{ bpm}$
Females: 95% CI: $\bar{x} \pm t_{\alpha/2}\frac{s}{\sqrt{n}} = 74.0 \pm 1.976 \cdot \frac{1.03}{\sqrt{147}} \Rightarrow 72.0 \text{ bpm} < \mu < 76.1 \text{ bpm}$
Although final conclusions about means of populations should not be based on the overlapping of confidence intervals, the intervals do not overlap, so adult females appear to have a mean pulse rate that is higher than the mean pulse rate of adult males.

27. The samples both appear to have a normal distribution.
McDonald's: 95% CI: $\bar{x} \pm t_{\alpha/2}\frac{s}{\sqrt{n}} = 179.3 \pm 2.010 \cdot \frac{62.94}{\sqrt{50}} \Rightarrow 161.4 \text{ sec} < \mu < 197.2 \text{ sec}$
Burger King: 95% CI: $\bar{x} \pm t_{\alpha/2}\frac{s}{\sqrt{n}} = 153.3 \pm 2.010 \cdot \frac{49.79}{\sqrt{50}} \Rightarrow 139.1 \text{ sec} < \mu < 167.5 \text{ sec}$
(Table: $139.2 \text{ sec} < \mu < 167.4 \text{ sec}$)
Although final conclusions about means of populations should not be based on the overlapping of confidence intervals, the intervals do overlap, so there does not appear to be a significant difference between the mean dinner service times at McDonald's and Burger King.

29. The sample size is $n = \left[\frac{z_{\alpha/2}\sigma}{E}\right]^2 = \left[\frac{2.575 \cdot 15}{4}\right]^2 = 94$. This does appear to be very practical.

31. The required sample size is $n = \left[\frac{z_{\alpha/2}\sigma}{E}\right]^2 = \left[\frac{1.96 \cdot 1.0}{0.01}\right]^2 = 38{,}416$ (Tech: 38,415). This does appear to be very practical.

33. The required sample size is $n = \left[\frac{z_{\alpha/2}\sigma}{E}\right]^2 = \left[\frac{1.96 \cdot 17.7}{0.5}\right]^2 = 4815$ (Tech: 4814). Yes, the assumption seems reasonable.

35. a. $\sigma \approx \frac{104-40}{4} = 16.0$; The required sample size is $n = \left[\frac{z_{\alpha/2}\sigma}{E}\right]^2 = \left[\frac{2.575 \cdot 16.0}{2}\right]^2 = 425$.

b. The required sample size is $n = \left[\frac{z_{\alpha/2}\sigma}{E}\right]^2 = \left[\frac{2.575 \cdot 11.3}{2}\right]^2 = 212$.

c. The result from part (a) is substantially larger than the result from part (b). The result from part (b) is likely to be better because it uses s instead of the estimated σ obtained from the range rule of thumb.

37. The sample size is greater than 30 and the data appear to be from a population that is normally distributed.

95% CI: $\bar{x} \pm z_{\alpha/2}\frac{\sigma}{\sqrt{n}} = 30.4 \pm 1.96 \cdot \frac{7.1}{\sqrt{205}} \Rightarrow 29.4 \text{ hg} < \mu < 31.4 \text{ hg}$

39. The second confidence interval is narrower, indicating that we have a more accurate estimate when the relatively large sample is from a relatively small finite population.

Large pop: 95% CI: $\bar{x} \pm t_{\alpha/2}\frac{s}{\sqrt{n}} = 0.8565 \pm 2.26 \cdot \frac{0.0518}{\sqrt{100}} \Rightarrow 0.8462 \text{ g} < \mu < 0.8668 \text{ g}$

Finite pop: 95% CI: $\bar{x} \pm t_{\alpha/2}\frac{s}{\sqrt{n}}\sqrt{\frac{N-n}{n-1}} = 0.8565 \pm 2.26 \cdot \frac{0.0518}{\sqrt{100}}\sqrt{\frac{465-100}{100-1}} \Rightarrow 0.8474 \text{ g} < \mu < 0.8656 \text{ g}$

Section 7-3: Estimating a Population Standard Deviation or Variance

1. $\sqrt{9027.8\left(\text{cm}^3\right)^2} < \sqrt{\sigma^2} < \sqrt{33299.8\left(\text{cm}^3\right)^2} \Rightarrow 95.0 \text{ cm}^3 < \sigma < 182.5 \text{ cm}^3$. We have 95% confidence that the limits of 95.0 cm^3 and 182.5 cm^3 contain the true value of the standard deviation of brain volumes.

3. The dotplot does not appear to depict sample data from a normally distributed population. The large sample size does not justify treating the values as being from a normally distributed population. Because the normality requirement is not satisfied, the confidence interval estimate of s should not be constructed using the methods of this section.

5. df = 24, $\chi_L^2 = 12.401$, and $\chi_R^2 = 39.364$

$$\sqrt{\frac{(n-1)s^2}{\chi_R^2}} < \sigma < \sqrt{\frac{(n-1)s^2}{\chi_L^2}}$$

$$\sqrt{\frac{(25-1)0.24^2}{39.364}} < \sigma < \sqrt{\frac{(25-1)0.24^2}{12.401}}$$

95% CI: $0.19 \text{ mg} < \sigma < 0.33 \text{ mg}$

7. df = 146, $\chi_L^2 = 105.761$, and $\chi_R^2 = 193.761$

(Table: $\chi_L^2 = 67.328$, and $\chi_R^2 = 140.169$)

$$\sqrt{\frac{(n-1)s^2}{\chi_R^2}} < \sigma < \sqrt{\frac{(n-1)s^2}{\chi_L^2}}$$

$$\sqrt{\frac{(147-1)65.4^2}{193.761}} < \sigma < \sqrt{\frac{(147-1)65.4^2}{105.741}}$$

95% CI: $56.8 < \sigma < 76.8$

(Table: $66.7 < \sigma < 96.3$)

9. df = 100, $\chi_L^2 = 74.222$, and $\chi_R^2 = 129.561$

$$\sqrt{\frac{(n-1)s^2}{\chi_R^2}} < \sigma < \sqrt{\frac{(n-1)s^2}{\chi_L^2}}$$

$$\sqrt{\frac{(106-1)0.62^2}{129.561}} < \sigma < \sqrt{\frac{(106-1)0.62^2}{74.222}}$$

95% CI: $0.56^\circ\text{F} < \sigma < 0.74^\circ\text{F}$

(Tech: $0.55^\circ\text{F} < \sigma < 0.72^\circ\text{F}$)

11. df = 15, $\chi_L^2 = 5.229$, and $\chi_R^2 = 30.578$

$$\sqrt{\frac{(n-1)s^2}{\chi_R^2}} < \sigma < \sqrt{\frac{(n-1)s^2}{\chi_L^2}}$$

$$\sqrt{\frac{(16-1)42.3^2}{30.578}} < \sigma < \sqrt{\frac{(16-1)42.3^2}{5.229}}$$

98% CI: $29.6 \text{ min} < \sigma < 71.6 \text{ min}$

No, the confidence interval does not indicate whether the treatment is effective.

13. The sample appears to have a normal distribution.
    df = 11, $\chi_L^2 = 4.575$, and $\chi_R^2 = 21.920$

$$\sqrt{\frac{(n-1)s^2}{\chi_R^2}} < \sigma < \sqrt{\frac{(n-1)s^2}{\chi_L^2}}$$

$$\sqrt{\frac{(12-1)2.216^2}{21.920}} < \sigma < \sqrt{\frac{(12-1)2.216^2}{4.575}}$$

95% CI: $1.6 < \sigma < 3.8$

15. The sample appears to have a normal distribution.
    df = 11, $\chi_L^2 = 3.816$, and $\chi_R^2 = 21.920$

$$\sqrt{\frac{(n-1)s^2}{\chi_R^2}} < \sigma < \sqrt{\frac{(n-1)s^2}{\chi_L^2}}$$

$$\sqrt{\frac{(12-1)4.08^2}{21.920}} < \sigma < \sqrt{\frac{(12-1)4.08^2}{3.816}}$$

95% CI: 2.9 mph $< \sigma <$ 6.9 mph

    Because traffic conditions vary considerably at different times during the day, the confidence interval is an estimate of the standard deviation of the population of speeds at 3:30 on a weekday, not other times.

17. a. The sample does not appear to have a normal distribution.
    df = 90, $\chi_L^2 = 65.647$, and $\chi_R^2 = 118.136$

$$\sqrt{\frac{(n-1)s^2}{\chi_R^2}} < \sigma < \sqrt{\frac{(n-1)s^2}{\chi_L^2}}$$

$$\sqrt{\frac{(93-1)0.5276^2}{118.136}} < \sigma < \sqrt{\frac{(93-1)0.5276^2}{65.647}}$$

95% CI: $0.47 < \sigma < 0.62$

(Table: $0.46 < \sigma < 0.62$)

    b. The sample does not appear to have a normal distribution.
    df = 90, $\chi_L^2 = 65.647$, and $\chi_R^2 = 118.136$

$$\sqrt{\frac{(n-1)s^2}{\chi_R^2}} < \sigma < \sqrt{\frac{(n-1)s^2}{\chi_L^2}}$$

$$\sqrt{\frac{(93-1)0.5608^2}{118.136}} < \sigma < \sqrt{\frac{(93-1)0.5608^2}{65.647}}$$

95% CI: $0.49 < \sigma < 0.66$

    c. The amounts of variation are about the same.

19. 19,205 is too large. There are not 19,205 statistics professors in the population, and even if there were, that sample size is too large to be practical.

21. The sample size is 48. No, with many very low incomes and a few high incomes, the distribution is likely to be skewed to the right and will not satisfy the requirement of a normal distribution.

23. $\chi_L^2 = \frac{1}{2}\left[-z_{\alpha/2} + \sqrt{2k-1}\right]^2 = \frac{1}{2}\left[-2.575 + \sqrt{2 \cdot 152 - 1}\right]^2 = 109.993$ and

$\chi_R^2 = \frac{1}{2}\left[z_{\alpha/2} + \sqrt{2k-1}\right]^2 = \frac{1}{2}\left[2.575 + \sqrt{2 \cdot 152 - 1}\right]^2 = 199.638$

(Tech: Using $z_{\alpha/2} = 2.575829303$, $\chi_L^2 = 109.980$ and $\chi_R^2 = 199.655$; The approximate values are quite close to the actual critical values.

Section 7-4: Bootstrapping: Using Technology for Estimates

1. Without replacement, every sample would be identical to the original sample, so the proportions or means or standard deviations or variances would all be the same, and there would be no confidence "interval."

3. Part (b): (There are only three elements, not five.), Part (d): (14 and 20 are not in the original sample), Part (e): (There are too many elements), are not possible bootstrap samples.

5. The proportions from the 10 samples (in ascending order) are: 0, 0, 0, 0, 0, 0.25, 0.25, 0.5, 0.5, and 0.5. $P_5 = 0.000$ and $P_{95} = 0.500$, so the 90% interval is $0.000 < p < 0.500$.

7. a. The means from the 10 samples (in ascending order) are: –0.25, 0.5, 0.5, 0.75, 3, 3, 3, 5, 8.25, and 9. $P_{10} = 0.125$ and $P_{90} = 8.625$, so the 80% interval is $0.1 \text{ kg} < \mu < 8.6 \text{ kg}$.
   b. The standard deviations from the 10 samples (in ascending order) are: 1.5, 2.363, 2.886751, 2.887, 4, 5.5, 5.715, 5.715, 5.715476, and 6.976. $P_{10} = 1.931$ and $P_{90} = 6.346$, so the 80% interval is $1.9 \text{ kg} < \sigma < 6.3 \text{ kg}$.

9. Answers will vary, but here are typical answers.
   a. 90% CI: $-0.8 \text{ kg} < \mu < 7.8 \text{ kg}$
   b. 90% CI: $1.2 \text{ kg} < \sigma < 7.0 \text{ kg}$

11. Answers will vary, but here are typical answers.
   a. 99% CI: $5.36 < \mu < 8.5$; This isn't dramatically different from $5.0 < \mu < 9.0$.
   b. 95% CI: $1.2 < \sigma < 2.9$; This isn't dramatically different from $1.6 < \sigma < 3.8$.

13. Answers will vary, but here is a typical result: 95% CI: $0.0608 < p < 0.123$. This is quite close to the confidence interval of $0.0615 < p < 0.121$ found in Exercise 13 from Section 7-1.

15. Answers will vary, but here is a typical result: 95% CI: $0.1356 < p < 0.152$. The result is essentially the same as the confidence interval of $0.135 < p < 0.152$ found in Exercise 15 from Section 7-1.

17. Answers will vary, but here is a typical result: 90% CI: $3.69 < \mu < 4.15$. This result is very close to the confidence interval $3.676 < \mu < 4.17$ found in Exercise 23 in Section 7-2.

19. a. Answers will vary, but here is a typical result: 95% CI: $233.6 \text{ sec} < \mu < 245.1 \text{ sec}$.
   b. 95% CI: $234.4 \text{ sec} < \mu < 246.0 \text{ sec}$
   c. The result from the bootstrap method is reasonably close to the result found using the methods of Section 7-2.

21. a. Answers will vary, but here is a typical result: 95% CI: $2.5 < \sigma < 3.3$.
   b. 95% CI: $2.4 < \sigma < 3.7$
   c. The confidence interval from the bootstrap method is not very different from the confidence interval found using the methods of Section 7-3. Because a histogram or normal quantile plot shows that the sample appears to be from a population not having a normal distribution, the bootstrap confidence interval of $2.5 < \sigma < 3.3$ would be a better estimate of σ.

23. Answers will vary, but here is a typical result using 10,000 bootstrap samples: $2.5 < \sigma < 3.3$. This result is the same as the confidence interval found using 1000 bootstrap samples. In this case, increasing the number of bootstrap samples from 1000 to 10,000 does not have much of an effect on the confidence interval.

Quick Quiz

1. $\hat{p} = \dfrac{0.692 + 0.748}{2} = 0.720$
2. We have 95% confidence that the limits of 0.692 and 0.748 contain the true value of the proportion of adults in the population who say that the law goes easy on celebrities.
3. $z_{0.01/2} = 2.576$ (Table: 2.575)
4. $40\% - 3.1\% < p < 40\% + 3.1\% \Rightarrow 36.9\% < p < 43.1\%$
5. $n = \dfrac{[z_{\alpha/2}]^2 \hat{p}\hat{q}}{E^2} = \dfrac{[1.96]^2 (0.25)}{0.04^2} = 601$
6. $n = \left[\dfrac{z_{\alpha/2}\sigma}{E}\right]^2 = \left[\dfrac{2.326 \cdot 15}{3}\right]^2 = 136$
7. There is a loose requirement that the sample values are from a normally distributed population.
8. The degrees of freedom is the number of sample values that can vary after restrictions have been imposed on all of the values. For the sample data described in Exercise 7, $\text{df} = 12 - 1 = 11$.
9. $t_{0.05/2} = 2.201$
10. No, the use of the χ^2 distribution has a fairly strict requirement that the data must be from a normal distribution. The bootstrap method could be used to find a 95% confidence interval estimate of σ.

Review Exercises

1. 95% CI: $\hat{p} \pm z_{\alpha/2}\sqrt{\dfrac{\hat{p}\hat{q}}{n}} = 0.40 \pm 1.96\sqrt{\dfrac{(0.40)(0.60)}{2036}} \Rightarrow 0.379 < p < 0.421$, or $37.9\% < p < 42.1\%$; Because we have 95% confidence that the limits of 37.9% and 42.1% contain the true percentage for the population of adults, we can safely say that fewer than 50% of adults prefer to get their news online.
2. $n = \dfrac{[z_{\alpha/2}]^2 \hat{p}\hat{q}}{E^2} = \dfrac{[1.645]^2 (0.25)}{0.04^2} = 423$
3. a. $\bar{x} = 2.926$

   b. 95% CI: $\bar{x} \pm t_{\alpha/2}\dfrac{s}{\sqrt{n}} = 2.926 \pm 2.201 \cdot \dfrac{0.278027}{\sqrt{12}} \Rightarrow 2.749 < \mu < 3.102$

   c. We have 95% confidence that the limits of 2.749 and 3.102 contain the value of the population mean μ.
4. $n = \left[\dfrac{z_{\alpha/2}\sigma}{E}\right]^2 = \left[\dfrac{2.575 \cdot 15}{4}\right]^2 = 94$
5. a. student *t* distribution

   b. normal distribution

   c. None of the three distributions is appropriate, but a confidence interval could be constructed by using bootstrap methods.

   d. χ^2 (chi-square distribution)

   e. normal distribution

6. a. $n = \frac{\left[z_{\alpha/2}\right]^2 \hat{p}\hat{q}}{E^2} = \frac{\left[1.96\right]^2 (0.25)}{0.03^2} = 1068$

   b. $n = \left[\frac{z_{\alpha/2}\sigma}{E}\right]^2 = \left[\frac{1.96 \cdot 17}{5}\right]^2 = 340$

   c. You must take the larger sample of 1068.

7. The sample appears to be normally distributed. 95% CI: $\bar{x} \pm t_{\alpha/2} \frac{s}{\sqrt{n}} = 143 \pm 2.201 \cdot \frac{259.78}{\sqrt{12}}$

   $\Rightarrow -22.1 \text{ sec} < \mu < 308.1 \text{ sec}.$

8. The sample appears to be normally distributed. df = 11, $\chi_L^2 = 3.816$, and $\chi_R^2 = 21.920$

   $$\sqrt{\frac{(n-1)s^2}{\chi_R^2}} < \sigma < \sqrt{\frac{(n-1)s^2}{\chi_L^2}}$$

   $$\sqrt{\frac{(12-1)259.78^2}{21.920}} < \sigma < \sqrt{\frac{(12-1)259.78^2}{3.816}}$$

   95% CI: $184.0 \text{ sec} < \sigma < 441.1 \text{ sec}$

9. Answers will vary, but here is a typical result: $7.1 \text{ sec} < \mu < 293.7 \text{ sec}.$

10. a. 95% CI: $\hat{p} \pm z_{\alpha/2}\sqrt{\frac{\hat{p}\hat{q}}{n}} = 0.02 \pm 1.96\sqrt{\frac{(0.02)(0.98)}{1000}} \Rightarrow 0.0113 < p < 0.0287$

    b. Answers will vary, but here is a typical result: $0.0120 < p < 0290.$

    c. The confidence intervals are quite close.

Cumulative Review Exercises

1. $\bar{x} = \frac{(-46)+(-32)+\cdots+(-21)+(-19)+\cdots+28+103}{12} = -3.6 \text{ min},\ Q_2 = \frac{(-23)+(-21)}{2} = -20.0 \text{ min},$

   $s = \sqrt{\frac{(-46-(-3.6))^2 + (-32-(-3.6))^2 + \cdots + (28-(-3.6))^2 + (103-(-3.6))^2}{12-1}} = 39.9 \text{ min},$

   range $= 103 - (-46) = 149.0$ min.

2. Using the range rule of thumb, the limit separating significantly low values is $\mu - 2\sigma = -3.6 - 2(39.9)$ $= -83.4$ min and the limit separating significantly high values is $\mu + 2\sigma = -3.6 + 2(39.9) = 76.2$ min. Because 103 min exceeds 76.2 min, the arrival delay time of 103 min is significantly high.

3. ratio level of measurement; continuous data.

4. The sample appears to not be normally distributed, so the confidence interval might not be a good estimate of the population mean. 95% CI: $\bar{x} \pm t_{\alpha/2} \frac{s}{\sqrt{n}} = -3.6 \pm 2.201 \cdot \frac{39.9}{\sqrt{12}} \Rightarrow -28.9 \text{ min} < \mu < 21.7 \text{ min}.$

5. a. $z_{x=15} = \frac{15.0-(-5.0)}{30.4} = 0.66$; which has a probability of $1 - 0.7454 = 0.02546$ (Tech: 0.553) to the right.

   b. The z score for the lower 75% is 0.67, which correspond to a time of $0.67 \cdot 30.4 + (-5.0) = 15.4$ min (Tech: 15.5 min).

6. $n = \left[\frac{z_{\alpha/2}\sigma}{E}\right]^2 = \left[\frac{1.96 \cdot 30.4}{5}\right]^2 = 143$ flights

7. 99% CI: $\hat{p} \pm z_{\alpha/2}\sqrt{\frac{\hat{p}\hat{q}}{n}} = 0.803 \pm 2.575\sqrt{\frac{(0.803)(0.197)}{1000}} \Rightarrow 0.771 < p < 0.835$, or $77.1\% < p < 83.5\%$

8. The graphs suggest that the population has a distribution that is skewed to the right instead of being normal. The histogram shows that some taxi-out times can be very long, and can occur with heavy traffic, but little or no traffic cannot make the taxi-out time very low. There is a minimum time required, regardless of traffic conditions. Construction of a confidence interval estimate of a population standard deviation has a fairly strict requirement that the sample data are from a normally distributed population, and the graphs show that this strict normality requirement is not satisfied.

Chapter 8: Hypothesis Testing

Section 8-1: Basics of Hypothesis Testing

1. Rejection of the claim about aspirin is more serious because it is a drug used for medical treatments. The wrong aspirin dosage could cause more serious adverse reactions than a wrong vitamin C dosage. It would be wise to use a smaller significance level for testing the claim about the aspirin.

3. a. H_0: $\mu = 174.1$ cm
   b. H_1: $\mu \neq 174.1$ cm
   c. Reject the null hypothesis or fail to reject the null hypothesis.
   d. No, in this case, the original claim becomes the null hypothesis. For the claim that the mean height of men is equal to 174.1 cm, we can either reject that claim or fail to reject it, but we cannot state that there is sufficient evidence to *support* that claim.

5. a. $p > 0.5$ (more than a majority)
   b. $H_0: p = 0.5$; $H_1: p > 0.5$

7. a. $\mu = 69$ bpm
   b. H_0: $\mu = 69$ bpm; H_1: $\mu \neq 69$ bpm

9. There is sufficient evidence to support the claim that most adults would erase all of their personal information online if they could.

11. There is not sufficient evidence to warrant rejection of the claim that the mean pulse rate (in beats per minute) of adult males is 69 bpm.

13. $z = \frac{\hat{p} - p}{\sqrt{\frac{pq}{n}}} = \frac{0.59 - 0.50}{\sqrt{\frac{(0.59)(0.41)}{565}}} = 4.28$ (if using $x = 0.59 \cdot 565 = 333$, $z = 4.25$)

15. $t = \frac{\bar{x} - \mu}{s/\sqrt{n}} = \frac{69.6 - 69}{11.3/\sqrt{153}} = 0.657$

17. a. right-tailed
    b. $P\text{-value} = P(z > 1.00) = 0.1587$
    c. $0.1587 > 0.05$; Fail to reject H_0.

19. a. two-tailed
    b. $P\text{-value} = 2 \cdot P(z > 2.01) = 0.0444$
    c. $0.0444 < 0.05$; Reject H_0.

21. a. Critical value: $z = 1.645$
    b. $1.00 < 1.645$; Fail to reject H_0.

22. b. $-2.01 < -1.645$; Reject H_0.

23. a. Critical values: $z = \pm 1.645$

24. a. Critical values: $z = \pm 1.96$
    b. $-1.96 < -1.94 < 1.96$; Fail to reject H_0.

25. a. $0.3257 > 0.05$; Fail to reject H_0.
    b. There is not sufficient evidence to support the claim that more than 58% of adults would erase all of their personal information online if they could.

27. a. $0.0095 < 0.05$; Reject H_0.
    b. There is sufficient evidence to warrant rejection of the claim that the mean pulse rate (in beats per minute) of adult males is 72 bpm.

29. Type I error: In reality $p = 0.1$, but we reject the claim that $p = 0.1$. Type II error: In reality $p \neq 0.1$, but we fail to reject the claim that $p = 0.1$.

31. Type I error: In reality $p = 0.87$, but we support the claim that $p > 0.87$. Type II error: In reality $p > 0.87$, but we fail to support that conclusion.

33. The power of 0.96 shows that there is a 96% chance of rejecting the null hypothesis of $p = 0.08$ when the true proportion is actually 0.18. That is, if the proportion of Chantix users who experience abdominal pain is actually 0.18, then there is a 96% chance of supporting the claim that the proportion of Chantix users who experience abdominal pain is greater than 0.08.

35. From $p = 0.5$, $\hat{p} = 0.5 + 1.645\sqrt{\frac{(0.5)(0.5)}{n}}$; from $p = 0.55$, since $P(z > -0.842) = 0.8000$,

$$\hat{p} = 0.55 - 0.842\sqrt{\frac{(0.55)(0.45)}{n}}$$

So: $0.5 + 1.645\sqrt{\frac{(0.5)(0.5)}{n}} = 0.55 - 0.842\sqrt{\frac{(0.55)(0.45)}{n}}$

$$0.5\sqrt{n} + 1.645\sqrt{0.25} = 0.55\sqrt{n} - 0.842\sqrt{0.2475}$$

$$0.05\sqrt{n} = 1.645\sqrt{0.25} + 0.842\sqrt{0.2475}$$

$$n = \left(\frac{1.645\sqrt{0.25} + 0.842\sqrt{0.2475}}{0.05}\right)^2 = 617$$

Section 8-2: Testing a Claim About a Proportion

1. a. $0.53 \cdot 510 = 270$
   b. $\hat{p} = 0.53$; The symbol $\hat{p}$ is used to represent a sample proportion.
3. The method based on a confidence interval is not equivalent to the *P*-value method and the critical value method.
5. a. left-tailed.
   b. $z = -4.46$
   c. *P*-value = 0.000004
   d. $H_0: p = 0.1$; Reject the null hypothesis.
   e. There is sufficient evidence to support the claim that less than 10% of treated subjects experience headaches.
7. a. two-tailed.
   b. $z = -1.69$
   c. *P*-value = 0.091
   d. $H_0: p = 0.92$; Fail to reject the null hypothesis.
   e. There is not sufficient evidence to warrant rejection of the claim that 92% of adults own cell phones.
9. $H_0: p = 0.10$; $H_1: p \neq 0.10$; Test statistic: $z = \frac{\frac{33}{362} - 0.10}{\sqrt{\frac{(0.10)(0.90)}{362}}} = -0.56$;

   P-value $= 2 \cdot P(z < -0.56) = 0.5755$ (Tech: 0.5751); Critical values: $z = \pm 1.96$;

   Fail to reject H_0. There is not sufficient evidence to warrant rejection of the claim that the rate of inaccurate orders is equal to 10%. With 10% of the orders being inaccurate, it appears that McDonald's should work to lower that rate.
11. $H_0: p = 0.5$; $H_1: p \neq 0.5$; Test statistic: $z = \frac{\frac{481}{882} - 0.5}{\sqrt{\frac{(0.5)(0.5)}{882}}} = 2.69$;

   P-value $= 2 \cdot P(z > 2.69) = 0.0072$ (Tech: 0.0071); Critical values: $z = \pm 2.575$ (Tech: $z = \pm 2.576$);

   Reject H_0. There is sufficient evidence to reject the claim that the proportion of those in favor is equal to 0.5. The result suggests that the politician is wrong in claiming that the responses are random guesses equivalent to a coin toss.
13. $H_0: p = 0.20$; $H_1: p > 0.20$; Test statistic: $z = \frac{\frac{52}{227} - 0.20}{\sqrt{\frac{(0.20)(0.80)}{227}}} = 1.10$;

   P-value $= P(z > 1.10) = 0.1357$ (Tech: 0.1367); Critical value: $z = 1.645$;

   Fail to reject H_0. There is not sufficient evidence to support the claim that more than 20% of OxyContin users develop nausea. However, with $\hat{p} = 0.229$, we see that a large percentage of OxyContin users experience nausea, so that rate does appear to be very high.

15. $H_0: p = 0.15$; $H_1: p < 0.15$; Test statistic: $z = \dfrac{\frac{717}{5000} - 0.15}{\sqrt{\frac{(0.15)(0.85)}{5000}}} = -1.31$;

P-value $= P(z < -1.31) = 0.1357$ (Tech: 0.1367); Critical value: $z = -2.33$;

Fail to reject H_0. There is not sufficient evidence to support the claim that the return rate is less than 15%.

17. $H_0: p = 0.512$; $H_1: p \neq 0.512$; Test statistic: $z = \dfrac{\frac{426}{860} - 0.512}{\sqrt{\frac{(0.512)(0.488)}{860}}} = -0.98$;

P-value $= 2 \cdot P(z < -0.98) = 0.3270$ (Tech: 0.3286); Critical values: $z = \pm 1.96$;

Fail to reject H_0. There is not sufficient evidence to warrant rejection of the claim that 51.2% of newborn babies are boys. The results do not *support* the belief that 51.2% of newborn babies are boys; the results merely show that there is not strong evidence against the rate of 51.2%.

19. $H_0: p = 0.80$; $H_1: p < 0.80$; Test statistic: $z = \dfrac{\frac{74}{98} - 0.80}{\sqrt{\frac{(0.80)(0.20)}{98}}} = -1.11$;

P-value $= P(z < -1.11) = 0.1335$ (Tech: 0.1332); Critical value: $z = -1.645$;

Fail to reject H_0. There is not sufficient evidence to support the claim that the polygraph results are correct less than 80% of the time. However, based on the sample proportion of correct results in 75.5% of the 98 cases, polygraph results do not appear to have the high degree of reliability that would justify the use of polygraph results in court, so polygraph test results should be prohibited as evidence in trials.

21. $H_0: p = 0.5$; $H_1: p \neq 0.5$; Test statistic: $z = \dfrac{\frac{123}{280} - 0.5}{\sqrt{\frac{(0.5)(0.5)}{280}}} = -2.03$;

P-value $= 2 \cdot P(z < -2.03) = 0.0424$ (Tech: 0.0422); Critical values: $z = \pm 1.645$;

Reject H_0. There is sufficient evidence to warrant rejection of the claim that touch therapists use a method equivalent to random guesses. However, their success rate of $123/280$, or 43.9%, indicates that they performed *worse* than random guesses, so they do not appear to be effective.

23. $H_0: p = 0.00034$; $H_0: p \neq 0.00034$; Test statistic: $z = \dfrac{\frac{135}{420{,}095} - 0.000340}{\sqrt{\frac{(0.000340)(0.99966)}{420{,}095}}} = -0.66$;

P-value $= 2 \cdot P(z < -0.66) = 0.5092$ (Tech: 0.5122); Critical values: $z = \pm 2.81$;

Fail to reject H_0. There is not sufficient evidence to support the claim that the rate is different from 0.0340%. Cell phone users should not be concerned about cancer of the brain or nervous system.

25. $H_0: p = 0.5$; $H_1: p > 0.5$; Test statistic: $z = \dfrac{\frac{28}{49} - 0.5}{\sqrt{\frac{(0.5)(0.5)}{49}}} = 1.00$;

P-value $= P(z > 1.00) = 0.1587$; Critical value: $z = 1.645$;

Fail to reject H_0. There is not sufficient evidence to support the claim that the probability of an NFC team Super Bowl win is greater than one-half.

27. $H_0: p = 0.5$; $H_1: p \neq 0.5$; Test statistic: $z = \dfrac{\frac{252}{460} - 0.5}{\sqrt{\frac{(0.5)(0.5)}{460}}} = 2.05$;

    P-value $= 2 \cdot P(z > 2.05) = 0.0404$ (Tech: 0.0402); Critical values: $z = \pm 1.96$;

    Reject H_0. There is sufficient evidence to warrant rejection of the claim that the coin toss is fair in the sense that neither team has an advantage by winning it. The coin toss rule does not appear to be fair. This helps explain why the overtime rules were changed.

29. $H_0: p = 0.5$; $H_1: p > 0.5$; Test statistic: $z = \dfrac{0.64 - 0.5}{\sqrt{\frac{(0.5)(0.5)}{21{,}346}}} = 40.91$ (using $x = 13{,}661, z = 40.90$);

    P-value $= P(z > 40.91) = 0.0001$ (Tech: 0.0000); Critical value: $z = 2.33$;

    Reject H_0. There is sufficient evidence to support the claim that most people believe that the Loch Ness monster exists. Because the sample is a voluntary-response sample, the conclusion about the population might not be valid.

31. $H_0: p = 0.791$; $H_1: p < 0.791$; Test statistic: $z = \dfrac{0.39 - 0.791}{\sqrt{\frac{(0.791)(0.209)}{870}}} = -29.09$ (using $x = 339, z = -29.11$);

    P-value $= P(z < -29.09) = 0.0001$ (Tech: 0.0000) (using $x = 339$, 0.3222, Tech: 0.3198);

    Critical value: $z = -2.33$;

    Reject H_0. There is sufficient evidence to support the claim that the percentage of selected Americans of Mexican ancestry is less than 79.1%, so the jury selection process appears to be biased.

33. $H_0: p = 0.5$; $H_1: p \neq 0.5$;

    Normal approximation:

    $$z = \frac{\frac{9}{10} - 0.5}{\sqrt{\frac{(0.5)(0.5)}{10}}} = 2.53; \quad P\text{-value} = 2 \cdot P(z > 2.53) = 0.0114$$

    Exact:

    $$P\text{-value} = 2 \cdot \left({}_{10}C_9 (0.5)^9 \left(0.5^1\right) + {}_{10}C_{10} (0.5)^{10} \left(0.5^0\right) \right) = 0.0215$$

    Continuity Correction:

    $$P\text{-value} = 2 \cdot \left({}_{10}C_9 (0.5)^9 \left(0.5^1\right) + {}_{10}C_{10} (0.5)^{10} \left(0.5^0\right) \right) - \frac{1}{2}\left({}_{10}C_9 (0.5)^9 \left(0.5^1\right) \right) = 0.0117$$

    $H_0: p = 0.4$; $H_1: p \neq 0.4$;

    Normal approximation:

    $$z = \frac{\frac{9}{10} - 0.4}{\sqrt{\frac{(0.4)(0.6)}{10}}} = 3.23; \quad P\text{-value} = 2 \cdot P(z > 3.23) = 0.0012$$

    Exact:

    $$P\text{-value} = 2 \cdot \left({}_{10}C_9 (0.4)^9 \left(0.6^1\right) + {}_{10}C_{10} (0.4)^{10} \left(0.6^0\right) \right) = 0.0034$$

33. (continued)

Continuity Correction:

$$P\text{-value} = 2 \cdot \left({}_{10}C_9 (0.4)^9 \left(0.6^1\right) + {}_{10}C_{10} (0.4)^{10} \left(0.6^0\right) \right) - \frac{1}{2} \left({}_{10}C_9 (0.4)^9 \left(0.6^1\right) \right) = 0.0018$$

H_0: $p = 0.5$; H_1: $p > 0.5$;

Normal approximation:

$$z = \frac{\frac{545}{1009} - 0.5}{\sqrt{\frac{(0.5)(0.5)}{1009}}} = 2.53; \quad P\text{-value} = P(z > 2.53) = 0.0057 \text{ (Tech: 0.0054)}$$

Exact:

$$P\text{-value} = {}_{1009}C_{545} (0.5)^{545} \left(0.5^{464}\right) + \cdots + {}_{1009}C_{1009} (0.5)^{1009} \left(0.5^0\right) = 0.0059$$

Continuity Correction:

$$P\text{-value} = {}_{1009}C_{545} (0.5)^{545} \left(0.5^{464}\right) + \cdots + {}_{1009}C_{1009} (0.5)^{1009} \left(0.5^0\right) - \frac{1}{2} \left({}_{1009}C_{545} (0.5)^{545} \left(0.5^{464}\right) \right)$$
$$= 0.0054$$

The P-values agree reasonably well with the large sample size of $n = 1009$. The normal approximation to the binomial distribution works better as the sample size increases.

35. a. From $p = 0.40$, $\hat{p} = 0.4 - 1.645\sqrt{\frac{(0.4)(0.6)}{50}} = 0.286$

From $p = 0.25$, $z = \frac{0.286 - 0.25}{\sqrt{\frac{(0.25)(0.75)}{50}}} = 0.588$; Power $= P(z < 0.588) = 0.7224$ (Tech: 0.7219)

b. $1 - 0.7224 = 0.2776$ (Tech: 0.2781)

c. The power of 0.7224 shows that there is a reasonably good chance of making the correct decision of rejecting the false null hypothesis. It would be better if the power were even higher, such as greater than 0.8 or 0.9.

Section 8-3: Testing a Claim About a Mean

1. The requirements are (1) the sample must be a simple random sample, and (2) either or both of these conditions must be satisfied: The population is normally distributed or $n > 30$. There is not enough information given to determine whether the sample is a simple random sample. Because the sample size is not greater than 30, we must check for normality, but the value of 583 sec appears to be an outlier, and a normal quantile plot or histogram suggests that the sample does not appear to be from a normally distributed population.

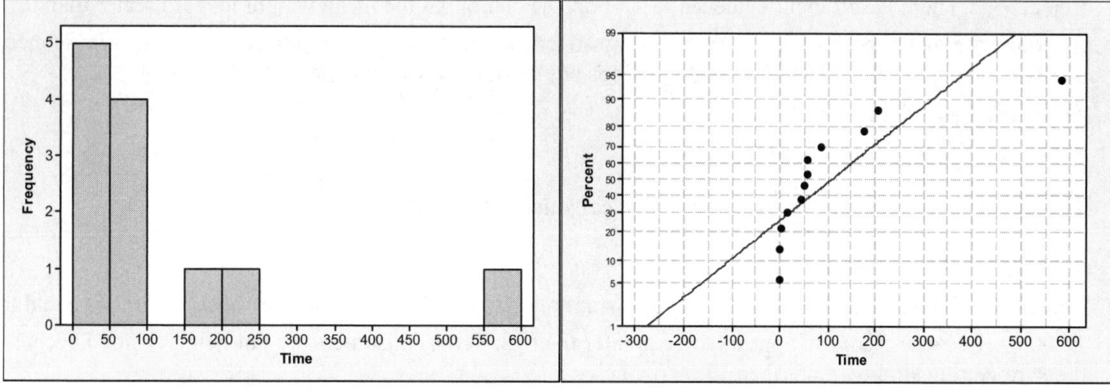

3. A t test is a hypothesis test that uses the Student t distribution, such as the method of testing a claim about a population mean as presented in this section. The t test methods are much more likely to be used than the z test methods because the t test does not require a known value of σ, and realistic hypothesis tests of claims about μ typically involve a population with an unknown value of σ.

5. P-value = 0.1301 (Table: $0.10 < P\text{-value} < 0.20$)
7. P-value = 0.2379 (Table: $P\text{-value} > 0.20$)
9. H_0: $\mu = 4.00$ Mbps; H_1: $\mu < 4.00$ Mbps; Test statistic: $t = -0.366$; P-value = 0.3579;
   Critical value (with $\alpha = 0.05$): $t = -1.667$ (Table: –1.676);
   Fail to reject H_0. There is not sufficient evidence to support the claim that the Sprint airport data speeds are from a population having a mean less than 4.00 Mbps.
11. H_0: $\mu = 0$ min; H_1: $\mu \neq 0$ min; Test statistic: $t = -8.720$; P-value = 0.0000;
   Critical value (with $\alpha = 0.05$): $t = \pm 1.970$;
   Reject H_0. There is sufficient evidence to warrant rejection of the claim that the mean prediction error is equal to zero. The predictions do not appear to be very accurate.
13. The data appear to follow a normal distribution and $n > 30$.
   H_0: $\mu = 4.00$; H_1: $\mu \neq 4.00$;
   Test statistic: $t = \frac{3.91 - 4.00}{0.53/\sqrt{93}} = -1.638$; Critical values: $t = \pm 1.987$ (Table: $t \approx \pm 1.676$);
   P-value = 0.1049 (Table: $P\text{-value} > 0.10$);
   Fail to reject H_0. There is not sufficient evidence to warrant rejection of the claim that the population of student course evaluations has a mean equal to 4.00.
15. The data cannot be verified to follow a normal distribution, but $n > 30$.
   H_0: $\mu = 0$; H_1: $\mu > 0$;
   Test statistic: $t = \frac{0.4 - 0}{21.0/\sqrt{49}} = -0.133$; Critical value: $t = 1.677$ (Table: $t \approx 1.676$);
   P-value = 0.4472 (Table: $P\text{-value} > 0.10$);
   Fail to reject H_0. There is not sufficient evidence to support the claim that with garlic treatment, the mean change in LDL cholesterol is greater than 0. The results suggest that the garlic treatment is not effective in reducing LDL cholesterol levels.
17. The data cannot be verified to follow a normal distribution, but $n > 30$.
   H_0: $\mu = 0$; H_1: $\mu > 0$;
   Test statistic: $t = \frac{3.0 - 0.0}{4.9/\sqrt{40}} = 3.872$; Critical value: $t = 2.426$;
   P-value = 0.0002 (Table: $P\text{-value} < 0.005$);
   Reject H_0. There is sufficient evidence to support the claim that the mean weight loss is greater than 0. Although the diet appears to have statistical significance, it does not appear to have practical significance, because the mean weight loss of only 3.0 lb does not seem to be worth the effort and cost.
19. The data appear right-skewed, but $n > 30$.
   H_0: $\mu = 12.00$ oz; H_1: $\mu \neq 12.00$ oz;
   Test statistic: $t = \frac{12.19 - 12.00}{0.11/\sqrt{36}} = 10.364$; Critical values: $t = \pm 2.030$;
   P-value = 0.0000 (Table: $P\text{-value} < 0.10$);
   Reject H_0. There is sufficient evidence to warrant rejection of the claim that the mean volume is equal to 12.00 oz. Because the mean appears to be greater than 12.00 oz, consumers are not being cheated because they are getting slightly more than 12.00 oz.

21. The sample data meet the loose requirement of having a normal distribution.
H_0: $\mu = 14$ μg/g; H_1: $\mu < 14$ μg/g;
Test statistic: $t = \frac{11.05 - 14.0}{6.46/\sqrt{10}} = -1.444$; Critical value: $t = -1.883$;
P-value = 0.0913 (Table: P-value > 0.05);
Fail to reject H_0. There is not sufficient evidence to support the claim that the mean lead concentration for all such medicines is less than 14 μg/g.

23. The data do not appear to follow a normal distribution and $n < 30$, so proceed with caution.
H_0: $\mu = 1000$ hic; H_1: $\mu < 1000$ hic;
Test statistic: $t = \frac{704 - 1000}{273/\sqrt{6}} = -2.661$; Critical value: $t = -3.365$;
P-value = 0.0224 (Table: 0.1 < P-value < 0.025);
Fail to reject H_0. There is not sufficient evidence to support the claim that the population mean is less than 1000 hic. There is not strong evidence that the mean is less than 1000 hic, and one of the booster seats has a measurement of 1210 hic, which does not satisfy the specified requirement of being less than 1000 hic.

25. The data appear to follow a normal distribution and $n > 30$.
H_0: $\mu = 75$ bpm; H_1: $\mu < 75$ bpm;
Test statistic: $t = \frac{74.04 - 75.00}{12.54/\sqrt{147}} = -0.927$; Critical value: $t = -1.655$ (Table: $t \approx -1.660$);
P-value = 0.1777 (Table: P-value > 0.10);
Fail to reject H_0. There is not sufficient evidence to support the claim that the mean pulse rate of adult females is less than 75 bpm.

27. The data appear to follow a normal distribution and $n > 30$.
H_0: $\mu = 90$ mm Hg; H_1: $\mu < 90$ mm Hg;
Test statistic: $t = \frac{70.163 - 90}{11.22/\sqrt{147}} = -21.435$; Critical value: $t = -1.655$ (Table: $t \approx -1.660$);
P-value = 0.0000 (Table: P-value < 0.005);
Reject H_0. There is sufficient evidence to support the claim that the adult female population has a mean diastolic blood pressure level less than 90 mm Hg. The conclusion addresses the mean of a population, not individuals, so we cannot conclude that there are no female adults in the sample with hypertension.

29. The data appear to follow a normal distribution and $n > 30$.
H_0: $\mu = 4.00$; H_1: $\mu \neq 4.00$;
Test statistic: $z = \frac{3.91 - 4.00}{0.53/\sqrt{93}} = -1.64$; Critical values: $z = \pm 1.96$;
P-value = 0.1015 (Table: 0.1010);
Fail to reject H_0. There is not sufficient evidence to warrant rejection of the claim that the population of student course evaluations has a mean equal to 4.00.
The null and alternative hypotheses are the same and the conclusions are the same. Results are not affected very much by the knowledge of σ.

31. $A = \frac{1.645(8 \cdot 499 + 3)}{8 \cdot 499 + 1} = 1.645823942$ and $t = \sqrt{499\left(e^{1.645823942^2/449} - 1\right)} = 1.648$; The approximation yields a critical t score of 1.648, which is the same as the value of 1.648 found from technology. The approximation appears to work quite well.

Section 8-4: Testing a Claim About a Standard Deviation of Variance

1. The sample must be a simple random sample and the sample must be from a normally distributed population. The normality requirement for a hypothesis test of a claim about a standard deviation is much more strict, meaning that the distribution of the population must be much closer to a normal distribution.

3. a. Reject H_0.
   b. Reject the claim that the new filling process results in volumes with the same standard deviation of 0.115 oz.
   c. It appears that with the new filling process, the variation among volumes has increased, so the volumes are not as consistent. The new filling process appears to be inferior to the original filling process.

5. H_0: $\sigma = 10$ bpm; H_1: $\sigma \neq 10$ bpm; Test statistic: $\chi^2 = 194.0888$; P-value $= 0.0239$;
   Reject H_0. There is sufficient evidence to warrant rejection of the claim that pulse rates of men have a standard deviation equal to 10 beats per minute. Using the normal range of 60 to 100 beats per minute is not very good for estimating σ in this case.

7. H_0: $\sigma = 2.08°\text{F}$; H_1: $\sigma < 2.08°\text{F}$; Test statistic: $\chi^2 = \frac{(n-1)s^2}{\sigma^2} = \frac{(106-1)0.62^2}{2.08^2} = 9.329$;
   P-value $= 0.0000$ (Table: P-value < 0.005); Critical value: $\chi^2 = 74.252$ (Table: $\chi^2 \approx 70.065$);
   Reject H_0. There is sufficient evidence to support the claim that body temperatures have a standard deviation less than 2.08°F. It is very highly unlikely that the conclusion in the hypothesis test in Example 5 from Section 8-3 would change because of a standard deviation from a different sample.

9. H_0: $\sigma = 27.8$ lb; H_1: $\sigma \neq 27.8$ lb; Test statistic: $\chi^2 = \frac{(n-1)s^2}{\sigma^2} = \frac{(20-1)18.6^2}{27.8^2} = 8.505$;
   P-value $= 0.0000$ (Table: P-value < 0.005); Critical values: $\chi^2 = 8.907$ and $\chi^2 = 32.852$;
   Reject H_0. There is sufficient evidence to warrant rejection of the claim that cans with thickness 0.0109 in. have axial loads with the same standard deviation as the axial loads of cans that are 0.0111 in. thick. The thickness of the cans does appear to affect the variation of the axial loads.

11. H_0: $\sigma = 0.15$ oz; H_1: $\sigma > 0.15$ oz; Test statistic: $\chi^2 = \frac{(n-1)s^2}{\sigma^2} = \frac{(27-1)0.17^2}{0.15^2} = 33.396$;
    P-value $= 0.1509$ (Table: P-value > 0.10); Critical value: $\chi^2 = 38.885$;
    Fail to reject H_0. There is not sufficient evidence to support the claim that the machine dispenses amounts with a standard deviation greater than the standard deviation of 0.15 oz specified in the machine design.

13. H_0: $\sigma = 32.2$ ft; H_1: $\sigma > 32.2$ ft; Test statistic: $\chi^2 = \frac{(n-1)s^2}{\sigma^2} = \frac{(12-1)52.441^2}{32.2^2} = 29.176$;
    P-value $= 0.0021$; Critical value: $\chi^2 = 19.675$;
    Reject H_0. There is sufficient evidence to support the claim that the new production method has errors with a standard deviation greater than 32.2 ft. The variation appears to be greater than in the past, so the new method appears to be worse, because there will be more altimeters that have larger errors. The company should take immediate action to reduce the variation.

15. H_0: $\sigma = 55.3$ sec; H_1: $\sigma \neq 55.3$ sec; Test statistic: $\chi^2 = \frac{(n-1)s^2}{\sigma^2} = \frac{(8-1)96.240^2}{55.93^2} = 27.726$;
    P-value $= 0.0084$ (Table: P-value < 0.01); Critical values: $\chi^2 = 0.989$ and $\chi^2 = 20.278$;
    Reject H_0. There is sufficient evidence to warrant rejection of the claim that service times at McDonald's have the same variation as service times at Wendy's. Drive-through service times during dinner times appear to vary more at McDonald's than those at Wendy's. Given the similar composition of the menus, McDonald's should consider methods for reducing the variation.

17. The data appear to be normally distributed. H_0: $\sigma = 55.93$ sec; H_1: $\sigma \neq 55.93$;

Test statistic: $\chi^2 = \frac{(n-1)s^2}{\sigma^2} = \frac{(50-1)62.938^2}{55.93^2} = 62.049$; P-value $= 0.1996$ (Table: P-value > 0.10);

Critical values: $\chi^2 = 27.249$ and $\chi^2 = 78.132$ (Table: $\chi^2 \approx 27.249$ and $\chi^2 \approx 78.132$);

Fail to reject H_0. There is not sufficient evidence to warrant rejection of the claim that service times at McDonald's have the same variation as service times at Wendy's. Drive-through service times during dinner times appear have about the same variation at McDonald's and Wendy's. No action is warranted.

19. Critical $\chi^2 = \frac{1}{2}\left(2.33 + \sqrt{2 \cdot 55 - 1}\right)^2 = 81.54$ (or 81.494 if using $z = 2.326348$ found from technology), which is reasonably close to the value of 22.465 obtained from STATDISK and Minitab.

Quick Quiz

1. a. t distribution
   b. Normal distribution
   c. Chi-square distribution
2. a. two-tailed
   b. left-tailed
   c. right-tailed
3. a. H_0: $p = 0.5$; H_1: $p > 0.5$
   b. Test statistic: $z = \frac{0.53 - 0.5}{\sqrt{\frac{(0.5)(0.5)}{532}}} = 1.38$ (using $x = 282$, $z = 1.39$);
   c. Fail to reject H_0.
   d. There is not sufficient evidence to support the claim that the majority of Internet users aged 18–29 use Instagram.
4. P-value $= 0.10$
5. true
6. false
7. false
8. No, all critical values of χ^2 are always positive.
9. The t test requires that the sample is from a normally distributed population, and the test is robust in the sense that the test works reasonably well if the departure from normality is not too extreme. The χ^2 (chi-square) test is not robust against a departure from normality, meaning that the test does not work well if the population has a distribution that is far from normal.
10. The only true statement is the one given in part (a).

Review Exercises

1. a. false
   b. true
   c. false
   d. false
   e. false
2. H_0: $p = 0.5$; H_1: $p > 0.5$; Test statistic: $z = \frac{\frac{40}{41} - 0.5}{\sqrt{\frac{(0.5)(0.5)}{41}}} = 6.09$;

   P-value $= P(z > 6.06) = 0.0001$ (Tech: 0.0000); Critical value: $z = 2.33$;

   Reject H_0. There is sufficient evidence to support the claim that the ballot selection method favors Democrats.

3. The data do not appear to follow a normal distribution, but $n > 30$.
   H_0: $\mu = 30$ years; H_1: $\mu > 30$ years;
   Test statistic: $t = \dfrac{36.2 - 30.0}{11.5/\sqrt{87}} = 5.029$; Critical value: $t = 2.370$ (Table: $t \approx 2.368$);
   P-value $= 0.0000$ (Table: P-value < 0.005);
   Reject H_0. There is sufficient evidence to support the claim that the mean age of actresses when they win Oscars is greater than 30 years.
4. The data cannot be verified to have a normal distribution, but $n > 30$.
   H_0: $\mu = 5.4$ million cells per microliter; H_1: $\mu < 5.4$ million cells per microliter;
   Test statistic: $t = \dfrac{4.932 - 5.4}{0.504/\sqrt{40}} = -5.873$; Critical value: $t = -2.426$;
   P-value $= 0.0000$ (Table: P-value < 0.005);
   Reject H_0. There is sufficient evidence to support the claim that the sample is from a population with a mean less than 5.4 million cells per microliter. The test deals with the distribution of sample means, not individual values, so the result does not suggest that each of the 40 males has a red blood cell count below 5.4 million cells per microliter.
5. H_0: $p = 0.43$; H_1: $p \neq 0.43$; Test statistic: $z = \dfrac{\frac{308}{611} - 0.43}{\sqrt{\frac{(0.43)(0.57)}{611}}} = 3.70$;
   P-value $= 2 \cdot P(z > 3.70) = 0.0002$; Critical values: $z = \pm 1.96$;
   Reject H_0. There is sufficient evidence to warrant rejection of the claim that the percentage who believe that they voted for the winning candidate is equal to 43%. There appears to be a substantial discrepancy between how people said that they voted and how they actually did vote.
6. The sample data meet the requirement of having a normal distribution.
   H_0: $\mu = 20.16$; H_1: $\mu < 20.16$;
   Test statistic: $t = \dfrac{18.76 - 20.16}{1.186/\sqrt{10}} = -3.732$; Critical value: $t = -2.821$;
   P-value $= 0.0023$ (Table: P-value < 0.005);
   Reject H_0. There is sufficient evidence to support the claim that the population of recent winners has a mean BMI less than 0.16. Recent winners appear to be significantly smaller than those from the 1920s and 1930s.
7. The sample data meet the requirement of having a normal distribution.
   H_0: $\sigma = 1.34$; H_1: $\sigma \neq 1.34$; Test statistic: $\chi^2 = \dfrac{(n-1)s^2}{\sigma^2} = \dfrac{(10-1)1.186^2}{1.34^2} = 7.053$;
   P-value $= 0.7368$ (Table: P-value > 0.20); Critical values: $\chi^2 = 1.735$ and $\chi^2 = 23.589$;
   Fail to reject H_0. There is not sufficient evidence to warrant rejection of the claim that the recent winners have BMI values with variation different from that of the 1920s and 1930s.
8. a. A type I error is the mistake of rejecting a null hypothesis when it is actually true. A type II error is the mistake of failing to reject a null hypothesis when in reality it is false.
   b. Type I error: In reality, the mean BMI is equal to 20.16, but we support the claim that the mean BMI is less than 20.16. Type II error: In reality, the mean BMI is less than 20.16, but we fail to support that claim.

Cumulative Review Exercises

1. a. $\bar{x} = \dfrac{23+26+27+28+29+32+34+38+43+44+45+48+51+51}{14} = 37.1 \text{ deaths}$

   b. $Q_2 = \dfrac{34+38}{2} = 36.0 \text{ deaths}$

   c. $s = \sqrt{\dfrac{(23-37.1)^2 + (26-37.1)^2 + \cdots + (51-37.1)^2 + (51-37.1)^2}{14-1}} = 9.8 \text{ deaths}$

   d. $s^2 = 9.8^2 = 96.8 \text{ deaths}^2$

   e. $\text{range} = 51 - 23 = 28.0 \text{ deaths}$

   f. The pattern of the data over time is not revealed by the statistics. A time-series graph would be very helpful in understanding the pattern over time.

2. a. ratio
   b. discrete
   c. quantitative
   d. No, the data are from recent and consecutive years, so they are not randomly selected.

3. 99% CI: $\bar{x} \pm t_{\alpha/2}\dfrac{s}{\sqrt{n}} = 37.1 \pm 3.012 \cdot \dfrac{9.84}{\sqrt{14}} \Rightarrow 29.1 \text{ deaths} < \mu < 45.0 \text{ deaths}$; We have 99% confidence that the limits of 29.1 deaths and 45.0 deaths contain the value of the population mean.

4. The sample data meet the loose requirement of having a normal distribution.
   H_0: $\mu = 72.6$ deaths; H_1: $\mu < 72.6$ deaths;

   Test statistic: $t = \dfrac{37.07 - 72.6}{9.84/\sqrt{14}} = -13.509$; Critical value: $t = -2.650$;

   P-value $= 0.0000$ (Table: P-value < 0.005);

   Reject H_0. There is sufficient evidence to support the claim that the mean number of annual lightning deaths is now less than the mean of 72.6 deaths from the 1980s. Possible factors: Shift in population from rural to urban areas; better lightning protection and grounding in electric and cable and phone lines; better medical treatment of people struck by lightning; fewer people use phones attached to cords; better weather predictions.

5. Because the vertical scale starts at 50 and not at 0, the difference between the number of males and the number of females is exaggerated, so the graph is deceptive by creating the false impression that males account for nearly all lightning strike deaths. A comparison of the numbers of deaths shows that the number of male deaths is roughly 4 times the number of female deaths, but the graph makes it appear that the number of male deaths is around 25 times the number of female deaths.

6. H_0: $p = 0.5$; H_1: $p > 0.5$; Test statistic: $z = \dfrac{\frac{232}{287} - 0.5}{\sqrt{\frac{(0.5)(0.5)}{287}}} = 10.45$;

   $P\text{-value} = P(z > 10.45) = 0.0001$ (Tech: 0.0000); Critical value: $z = 2.33$;

   Reject H_0. There is sufficient evidence to support the claim that the proportion of male deaths is greater than $1/2$. More males are involved in certain outdoor activities such as construction, fishing, and golf.

7. 95% CI: $\hat{p} \pm z_{\alpha/2}\sqrt{\dfrac{\hat{p}\hat{q}}{n}} = \dfrac{232}{287} \pm 1.96\sqrt{\dfrac{\left(\frac{232}{287}\right)\left(\frac{55}{287}\right)}{287}} \Rightarrow 0.762 < p < 0.854$; Because the entire confidence interval is greater than 0.5, it does not seem feasible that males and females have equal chances of being killed by lightning.

8. a. $0.8 \cdot 0.8 \cdot 0.8 = 0.512$
   b. $0.2 \cdot 0.2 \cdot 0.2 = 0.008$
   c. $1 - 0.2 \cdot 0.2 \cdot 0.2 = 0.992$
   d. ${}_5C_3(0.8)^3(0.2^2) = 0.205$
   e. $\mu = np = 50 \cdot 0.8 = 40$ males; $\sigma = \sqrt{npq} = \sqrt{50 \cdot 0.8 \cdot 0.2} = 2.8$ males
   f. Yes, using the range rule of thumb, values above $\mu + 2\sigma = 40.0 + 2(2.8) = 45.6$ are considered significantly high. Since 46 is greater than 45.6, 46 males would be a significantly high number in a group of 50.

Chapter 9: Inferences from Two Samples

Section 9-1: Two Proportions

1. The samples are simple random samples that are independent. For each of the two groups, the number of successes is at least 5 and the number of failures is at least 5. (Depending on what we call a success, the four numbers are 33, 115, 201,229 and 200,745 and all of those numbers are at least 5.) The requirements are satisfied.

3. a. $H_0: p_1 = p_2; H_1: p_1 < p_2$

   b. There is sufficient evidence to support the claim that the rate of polio is less for children given the Salk vaccine than it is for children given a placebo. The Salk vaccine appears to be effective.

5. $H_0: p_1 = p_2; H_1: p_1 > p_2$; population$_1$ = vinyl gloves, population$_2$ = latex gloves;
   Test statistic: $z = 12.82$; P-value = 0.0000; Critical value: $z = 2.33$; Reject H_0. There is sufficient evidence to support the claim that vinyl gloves have a greater virus leak rate than latex gloves.

For Exercises 7–21, assume that the data fit the requirements for the statistical methods for two proportions unless otherwise indicated.

7. a. $H_0: p_1 = p_2; H_1: p_1 > p_2$; population$_1$ = cars, population$_2$ = trucks;
   Test statistic: $z = -0.95$; P-value = 0.8280 (Table: 0.5289); Critical value: $z = 1.645$; Fail to reject H_0. There is not sufficient evidence to support the claim that car owners violate license plate laws at a higher rate than owners of commercial trucks.

$$\bar{p} = \frac{239+45}{2049+334} = \frac{284}{2383}; \bar{q} = 1 - \frac{284}{2383} = \frac{2099}{2383};$$

$$z = \frac{(\hat{p}_1 - \hat{p}_2) - (p_1 - p_2)}{\sqrt{\frac{\bar{p}\bar{q}}{n_1} + \frac{\bar{p}\bar{q}}{n_2}}} = \frac{\left(\frac{239}{2049} - \frac{45}{334}\right) - 0}{\sqrt{\frac{\left(\frac{284}{2383}\right)\left(\frac{2099}{2383}\right)}{2049} + \frac{\left(\frac{284}{2383}\right)\left(\frac{2099}{2383}\right)}{334}}} = -0.95$$

   b. 90% CI: $-0.0510 < p_1 - p_2 < 0.0148$; Because the confidence interval limits contain 0, there is not a significant difference between the two proportions. There is not sufficient evidence to support the claim that car owners violate license plate laws at a higher rate than owners of commercial trucks.

$$(\hat{p}_1 - \hat{p}_2) \pm z_{\alpha/2}\sqrt{\frac{\hat{p}_1\hat{q}_1}{n_1} + \frac{\hat{p}_2\hat{q}_2}{n_2}} = \left(\frac{239}{2049} - \frac{45}{334}\right) \pm 1.645\sqrt{\frac{\left(\frac{239}{2049}\right)\left(\frac{1810}{2049}\right)}{2049} + \frac{\left(\frac{45}{334}\right)\left(\frac{289}{334}\right)}{334}}$$

9. a. $H_0: p_1 = p_2; H_1: p_1 > p_2$; population$_1$ = sustained care, population$_2$ = standard care;
   Test statistic: $z = 2.64$; P-value = 0.0041; Critical value: $z = 2.33$; Reject H_0. There is sufficient evidence to support the claim that the rate of success for smoking cessation is greater with the sustained care program.

$$\bar{p} = \frac{51+30}{198+199} = \frac{81}{397}; \bar{q} = 1 - \frac{81}{397} = \frac{316}{397};$$

$$z = \frac{(\hat{p}_1 - \hat{p}_2) - (p_1 - p_2)}{\sqrt{\frac{\bar{p}\bar{q}}{n_1} + \frac{\bar{p}\bar{q}}{n_2}}} = \frac{\left(\frac{51}{198} - \frac{30}{199}\right) - 0}{\sqrt{\frac{\left(\frac{81}{397}\right)\left(\frac{316}{397}\right)}{198} + \frac{\left(\frac{81}{397}\right)\left(\frac{316}{397}\right)}{199}}} = 2.64$$

9. (continued)

   b. 98% CI: $0.0135 < p_1 - p_2 < 0.200$ (Table: $0.0134 < p_1 - p_2 < 0.200$); Because the confidence interval limits do not contain 0, there is a significant difference between the two proportions. Because the interval consists of positive numbers only, it appears that the success rate for the sustained care program is greater than the success rate for the standard care program.

$$(\hat{p}_1 - \hat{p}_2) \pm z_{\alpha/2}\sqrt{\frac{\hat{p}_1\hat{q}_1}{n_1} + \frac{\hat{p}_2\hat{q}_2}{n_2}} = \left(\frac{51}{198} - \frac{30}{199}\right) \pm 2.33\sqrt{\frac{\left(\frac{51}{198}\right)\left(\frac{147}{198}\right)}{198} + \frac{\left(\frac{30}{199}\right)\left(\frac{169}{199}\right)}{199}}$$

   c. Based on the samples, the success rates of the programs are 25.8% (sustained care) and 15.1% (standard care). That difference does appear to be substantial, so the difference between the programs does appear to have practical significance.

11. a. $H_0: p_1 = p_2$; $H_1: p_1 > p_2$; population$_1$ = over age 55, population$_2$ = under age 25; Test statistic: $z = 6.44$; P-value = 0.0000 (Table: 0.0001); Critical value: $z = 2.33$; Reject H_0. There is sufficient evidence to support the claim that the proportion of people over 55 who dream in black and white is greater than the proportion of those under 25.

$$\bar{p} = \frac{68+13}{306+298} = \frac{81}{604}; \bar{q} = 1 - \frac{81}{604} = \frac{523}{604};$$

$$z = \frac{(\hat{p}_1 - \hat{p}_2) - (p_1 - p_2)}{\sqrt{\frac{\bar{p}\bar{q}}{n_1} + \frac{\bar{p}\bar{q}}{n_2}}} = \frac{\left(\frac{68}{306} - \frac{13}{298}\right) - 0}{\sqrt{\frac{\left(\frac{81}{604}\right)\left(\frac{523}{604}\right)}{306} + \frac{\left(\frac{81}{604}\right)\left(\frac{523}{604}\right)}{298}}} = 6.44$$

   b. 98% CI: $0.117 < p_1 - p_2 < 0.240$; Because the confidence interval limits do not include 0, it appears that the two proportions are not equal. Because the confidence interval limits include only positive values, it appears that the proportion of people over 55 who dream in black and white is greater than the proportion of those under 25.

$$(\hat{p}_1 - \hat{p}_2) \pm z_{\alpha/2}\sqrt{\frac{\hat{p}_1\hat{q}_1}{n_1} + \frac{\hat{p}_2\hat{q}_2}{n_2}} = \left(\frac{68}{306} - \frac{13}{298}\right) \pm 2.33\sqrt{\frac{\left(\frac{68}{306}\right)\left(\frac{238}{306}\right)}{306} + \frac{\left(\frac{13}{298}\right)\left(\frac{285}{298}\right)}{298}}$$

   c. The results suggest that the proportion of people over 55 who dream in black and white is greater than the proportion of those under 25, but the results cannot be used to verify the cause of that difference.

13. a. $H_0: p_1 = p_2$; $H_1: p_1 > p_2$; population$_1$ = wearing seatbelt, population$_2$ = not wearing seatbelt; Test statistic: $z = 6.11$; P-value = 0.0000 (Table: 0.0001); Critical value: $z = 1.645$; Reject H_0. There is sufficient evidence to support the claim that the fatality rate is higher for those not wearing seat belts.

$$\bar{p} = \frac{31+16}{2823+7765} = \frac{47}{10{,}588}; \bar{q} = 1 - \frac{47}{10{,}588} = \frac{10{,}541}{10{,}588};$$

$$z = \frac{(\hat{p}_1 - \hat{p}_2) - (p_1 - p_2)}{\sqrt{\frac{\bar{p}\bar{q}}{n_1} + \frac{\bar{p}\bar{q}}{n_2}}} = \frac{\left(\frac{31}{2823} - \frac{16}{7765}\right) - 0}{\sqrt{\frac{\left(\frac{47}{10{,}588}\right)\left(\frac{10{,}541}{10{,}588}\right)}{2823} + \frac{\left(\frac{47}{10{,}588}\right)\left(\frac{10{,}541}{10{,}588}\right)}{7765}}} = 6.11$$

13. (continued)

b. 90% CI: $0.00559 < p_1 - p_2 < 0.0123$; Because the confidence interval limits do not include 0, it appears that the two fatality rates are not equal. Because the confidence interval limits include only positive values, it appears that the fatality rate is higher for those not wearing seat belts.

$$(\hat{p}_1 - \hat{p}_2) \pm z_{\alpha/2}\sqrt{\frac{\hat{p}_1\hat{q}_1}{n_1} + \frac{\hat{p}_2\hat{q}_2}{n_2}} = \left(\frac{31}{2823} - \frac{16}{7765}\right) \pm 1.645\sqrt{\frac{\left(\frac{31}{2823}\right)\left(\frac{2972}{2823}\right)}{2823} + \frac{\left(\frac{16}{7765}\right)\left(\frac{7749}{7765}\right)}{7765}}$$

c. The results suggest that the use of seat belts is associated with fatality rates lower than those associated with not using seat belts.

15. a. $H_0: p_1 = p_2$; $H_1: p_1 \neq p_2$; population$_1$ = echinacea, population$_2$ = placebo;

Test statistic: $z = 0.57$; P-value = 0.5720 (Table: 0.5868); Critical values: $z = \pm 1.96$; Fail to reject H_0. There is not sufficient evidence to support the claim that Echinacea treatment has an effect.

$$\bar{p} = \frac{40+88}{45+103} = \frac{32}{37};\ \bar{q} = 1 - \frac{32}{37} = \frac{5}{37};$$

$$z = \frac{(\hat{p}_1 - \hat{p}_2) - (p_1 - p_2)}{\sqrt{\frac{\bar{p}\bar{q}}{n_1} + \frac{\bar{p}\bar{q}}{n_2}}} = \frac{\left(\frac{40}{45} - \frac{88}{103}\right) - 0}{\sqrt{\frac{\left(\frac{32}{37}\right)\left(\frac{5}{37}\right)}{45} + \frac{\left(\frac{32}{37}\right)\left(\frac{5}{37}\right)}{103}}} = 0.57$$

b. 95% CI: $-0.0798 < p_1 - p_2 < 0.149$; Because the confidence interval limits do contain 0, there is not a significant difference between the two proportions. There is not sufficient evidence to support the claim that Echinacea treatment has an effect.

$$(\hat{p}_1 - \hat{p}_2) \pm z_{\alpha/2}\sqrt{\frac{\hat{p}_1\hat{q}_1}{n_1} + \frac{\hat{p}_2\hat{q}_2}{n_2}} = \left(\frac{40}{45} - \frac{88}{103}\right) \pm 1.96\sqrt{\frac{\left(\frac{40}{45}\right)\left(\frac{5}{45}\right)}{45} + \frac{\left(\frac{88}{103}\right)\left(\frac{15}{103}\right)}{103}}$$

c. Echinacea does not appear to have a significant effect on the infection rate. Because it does not appear to have an effect, it should not be recommended.

17. a. $H_0: p_1 = p_2$; $H_1: p_1 < p_2$; population$_1$ = used left ear, population$_2$ = used right ear;

Test statistic: $z = -7.94$; P-value = 0.0000 (Table: 0.0001); Critical value: $z = -2.33$; Reject H_0. There is sufficient evidence to support the claim that the rate of right-handedness for those who prefer to use their left ear for cell phones is less than the rate of right-handedness for those who prefer to use their right ear for cell phones.

$$\bar{p} = \frac{166+436}{216+452} = \frac{301}{334};\ \bar{q} = 1 - \frac{301}{334} = \frac{33}{334};$$

$$z = \frac{(\hat{p}_1 - \hat{p}_2) - (p_1 - p_2)}{\sqrt{\frac{\bar{p}\bar{q}}{n_1} + \frac{\bar{p}\bar{q}}{n_2}}} = \frac{\left(\frac{166}{216} - \frac{436}{452}\right) - 0}{\sqrt{\frac{\left(\frac{301}{334}\right)\left(\frac{33}{334}\right)}{216} + \frac{\left(\frac{301}{334}\right)\left(\frac{33}{334}\right)}{452}}} = -7.94$$

b. 98% CI: $-0.266 < p_1 - p_2 < -0.126$; Because the confidence interval limits do not contain 0, there is a significant difference between the two proportions. Because the interval consists of negative numbers only, it appears that the claim is supported. The difference between the populations does appear to have practical significance.

$$(\hat{p}_1 - \hat{p}_2) \pm z_{\alpha/2}\sqrt{\frac{\hat{p}_1\hat{q}_1}{n_1} + \frac{\hat{p}_2\hat{q}_2}{n_2}} = \left(\frac{166}{216} - \frac{436}{452}\right) \pm 2.33\sqrt{\frac{\left(\frac{166}{216}\right)\left(\frac{50}{216}\right)}{216} + \frac{\left(\frac{436}{452}\right)\left(\frac{16}{452}\right)}{452}}$$

19. a. $H_0: p_1 = p_2$; $H_1: p_1 > p_2$; population$_1$ = oxygen, population$_2$ = placebo;
Test statistic: $z = 9.97$; P-value = 0.0000 (Table: 0.0001); Critical value: $z = 2.33$; Reject H_0. There is sufficient evidence to support the claim that the cure rate with oxygen treatment is higher than the cure rate for those given a placebo. It appears that the oxygen treatment is effective.

$$\bar{p} = \frac{116+29}{150+148} = \frac{145}{298}; \bar{q} = 1 - \frac{145}{298} = \frac{153}{298};$$

$$z = \frac{(\hat{p}_1 - \hat{p}_2) - (p_1 - p_2)}{\sqrt{\frac{\bar{p}\bar{q}}{n_1} + \frac{\bar{p}\bar{q}}{n_2}}} = \frac{\left(\frac{116}{150} - \frac{29}{148}\right) - 0}{\sqrt{\frac{\left(\frac{145}{298}\right)\left(\frac{153}{298}\right)}{150} + \frac{\left(\frac{145}{298}\right)\left(\frac{153}{298}\right)}{148}}} = 9.97$$

b. 98% CI: $0.467 < p_1 - p_2 < 0.687$; Because the confidence interval limits do not include 0, it appears that the two cure rates are not equal. Because the confidence interval limits include only positive values, it appears that the cure rate with oxygen treatment is higher than the cure rate for those given a placebo. It appears that the oxygen treatment is effective.

$$(\hat{p}_1 - \hat{p}_2) \pm z_{\alpha/2}\sqrt{\frac{\hat{p}_1\hat{q}_1}{n_1} + \frac{\hat{p}_2\hat{q}_2}{n_2}} = \left(\frac{116}{150} - \frac{29}{148}\right) \pm 2.33\sqrt{\frac{\left(\frac{116}{150}\right)\left(\frac{34}{150}\right)}{150} + \frac{\left(\frac{29}{148}\right)\left(\frac{119}{148}\right)}{148}}$$

c. The results suggest that the oxygen treatment is effective in curing cluster headaches.

21. a. $H_0: p_1 = p_2$; $H_1: p_1 < p_2$; population$_1$ = male, population$_2$ = female;
Test statistic: $z = -1.17$; P-value = 0.1214 (Table: 0.1210); Critical value: $z = -2.33$; Fail to reject H_0. There is not sufficient evidence to support the claim that the rate of left-handedness among males is less than that among females.

$$\bar{p} = \frac{23+65}{240+520} = \frac{11}{95}; \bar{q} = 1 - \frac{11}{95} = \frac{84}{95};$$

$$z = \frac{(\hat{p}_1 - \hat{p}_2) - (p_1 - p_2)}{\sqrt{\frac{\bar{p}\bar{q}}{n_1} + \frac{\bar{p}\bar{q}}{n_2}}} = \frac{\left(\frac{23}{240} - \frac{65}{520}\right) - 0}{\sqrt{\frac{\left(\frac{11}{95}\right)\left(\frac{84}{95}\right)}{240} + \frac{\left(\frac{11}{95}\right)\left(\frac{84}{95}\right)}{520}}} = -1.17$$

b. 98% CI: $-0.0848 < p_1 - p_2 < -0.0264$ (Table: $-0.0849 < p_1 - p_2 < -0.0265$); Because the confidence interval limits include 0, there does not appear to be a significant difference between the rate of left-handedness among males and the rate among females. There is not sufficient evidence to support the claim that the rate of left-handedness among males is less than that among females.

$$(\hat{p}_1 - \hat{p}_2) \pm z_{\alpha/2}\sqrt{\frac{\hat{p}_1\hat{q}_1}{n_1} + \frac{\hat{p}_2\hat{q}_2}{n_2}} = \left(\frac{23}{240} - \frac{65}{520}\right) \pm 2.33\sqrt{\frac{\left(\frac{23}{240}\right)\left(\frac{217}{240}\right)}{240} + \frac{\left(\frac{65}{520}\right)\left(\frac{455}{520}\right)}{520}}$$

c. The rate of left-handedness among males does not appear to be less than the rate of left-handedness among females.

23. $n=\frac{z_{\alpha/2}^2}{2E^2}=\frac{1.96^2}{2\cdot 0.03^2}=2135$; The samples should include 2135 men and 2135 women.

25. a. 95% CI. $0.0227 < p_1 - p_2 < 0.217$; population$_1$ = first sample, population$_2$ = second sample; Because the confidence interval limits do not contain 0, it appears that $p_1 = p_2$ can be rejected.

$$(\hat{p}_1-\hat{p}_2)\pm z_{\alpha/2}\sqrt{\frac{\hat{p}_1\hat{q}_1}{n_1}+\frac{\hat{p}_2\hat{q}_2}{n_2}}=\left(\frac{112}{200}-\frac{88}{200}\right)\pm 1.96\sqrt{\frac{\left(\frac{112}{200}\right)\left(\frac{88}{200}\right)}{200}+\frac{\left(\frac{88}{200}\right)\left(\frac{112}{200}\right)}{200}}$$

b. First sample: 95% CI: $\hat{p}\pm z_{\alpha/2}\sqrt{\frac{\hat{p}\hat{q}}{n}}=\frac{112}{200}\pm 1.96\sqrt{\frac{\left(\frac{112}{200}\right)\left(\frac{88}{200}\right)}{200}}\Rightarrow 0.491 < p_1 < 0.629$

Second sample: 95% CI: $\hat{p}\pm z_{\alpha/2}\sqrt{\frac{\hat{p}\hat{q}}{n}}=\frac{88}{200}\pm 1.96\sqrt{\frac{\left(\frac{88}{200}\right)\left(\frac{112}{200}\right)}{200}}\Rightarrow 0.371 < p_2 < 0.509$

Because the confidence intervals do overlap, it appears that $p_1 = p_2$ cannot be rejected.

c. $H_0\colon p_1 = p_2$; $H_1\colon p_1 \neq p_2$; population$_1$ = first sample, population$_2$ = second sample; Test statistic: $z = 2.40$; P-value = 0.0164; Critical values: $z = \pm 1.96$; Reject H_0. There is sufficient evidence to reject $p_1 = p_2$.

$$\bar{p}=\frac{112+88}{200+200}=\frac{1}{2};\ \bar{q}=1-\frac{1}{2}=\frac{1}{2};$$

$$z=\frac{(\hat{p}_1-\hat{p}_2)-(p_1-p_2)}{\sqrt{\frac{\bar{p}\bar{q}}{n_1}+\frac{\bar{p}\bar{q}}{n_2}}}=\frac{\left(\frac{112}{200}-\frac{88}{200}\right)-0}{\sqrt{\frac{\left(\frac{1}{2}\right)\left(\frac{1}{2}\right)}{200}+\frac{\left(\frac{1}{2}\right)\left(\frac{1}{2}\right)}{200}}}=2.40$$

d. Reject $p_1 = p_2$. The least effective method is using the overlap between the individual confidence intervals.

Section 9-2: Two Means: Independent Samples

1. Only part (c) describes independent samples.

3. a. yes
   b. yes
   c. 98%

5. a. $H_0\colon \mu_1 = \mu_2$; $H_1\colon \mu_1 < \mu_2$; population$_1$ = Diet Coke, population$_2$ = regular Coke; Test statistic: $t = -22.092$; P-value = 0.0000 (Table: P-value < 0.005); Critical value: $t = -1.672$ (Table: $t = -1.690$); Reject H_0. There is sufficient evidence to support the claim that the contents of cans of Diet Coke have weights with a mean that is less than the mean for regular Coke.

$$t=\frac{(\bar{x}_1-\bar{x}_2)-(\mu_1-\mu_2)}{\sqrt{\frac{s_1^2}{n_1}+\frac{s_2^2}{n_2}}}=\frac{(0.78479-0.81682)-0}{\sqrt{\frac{0.00439^2}{36}+\frac{0.00751^2}{36}}}=-22.092\ (\text{df}=35)$$

b. 90% CI: $-0.03445 \text{ lb} < \mu_1 - \mu_2 < -0.02961 \text{ lb}$ (Table: $-0.03448 \text{ lb} < \mu_1 - \mu_2 < -0.02958 \text{ lb}$);

$$(\bar{x}_1-\bar{x}_2)\pm t_{\alpha/2}\sqrt{\frac{s_1^2}{n_1}+\frac{s_2^2}{n_2}}=(0.78479-0.81682)\pm 1.690\sqrt{\frac{0.00439^2}{36}+\frac{0.00751^2}{36}}\ (\text{df}=35)$$

c. The contents in cans of Diet Coke appear to weigh less, probably due to the sugar present in regular Coke but not Diet Coke.

7. a. H_0: $\mu_1 = \mu_2$; H_1: $\mu_1 < \mu_2$; population$_1$ = red background, population$_2$ = red background;
Test statistic: $t = -2.979$; P-value = 0.0021 (Table: P-value < 0.005); Critical value: $t = -2.392$ (Table: $t = -2.441$); Reject H_0. There is sufficient evidence to support the claim that blue enhances performance on a creative task.

$$t = \frac{(\bar{x}_1 - \bar{x}_2) - (\mu_1 - \mu_2)}{\sqrt{\frac{s_1^2}{n_1} + \frac{s_2^2}{n_2}}} = \frac{(3.39 - 3.97) - 0}{\sqrt{\frac{0.97^2}{35} + \frac{0.63^2}{36}}} = -2.979 \text{ (df = 34)}$$

b. 98% CI: $-1.05 < \mu_1 - \mu_2 < -0.11$ (Table: $-1.06 < \mu_1 - \mu_2 < -0.10$); The confidence interval consists of negative numbers only and does not include 0, so the mean creativity score with the red background appears to be less than the mean creativity score with the blue background. It appears that blue enhances performance on a creative task.

$$(\bar{x}_1 - \bar{x}_2) \pm t_{\alpha/2}\sqrt{\frac{s_1^2}{n_1} + \frac{s_2^2}{n_2}} = (3.39 - 3.97) \pm 2.41\sqrt{\frac{0.97^2}{35} + \frac{0.63^2}{36}} \text{ (df = 34)}$$

9. a. H_0: $\mu_1 = \mu_2$; H_1: $\mu_1 > \mu_2$; population$_1$ = magnet treatment, population$_2$ = sham treatment;
Test statistic: $t = 0.132$; P-value = 0.4480 (Table: P-value > 0.10); Critical value: $t = 1.691$ (Table: $t = 1.729$); Fail to reject H_0. There is not sufficient evidence to support the claim that the magnets are effective in reducing pain.

$$t = \frac{(\bar{x}_1 - \bar{x}_2) - (\mu_1 - \mu_2)}{\sqrt{\frac{s_1^2}{n_1} + \frac{s_2^2}{n_2}}} = \frac{(0.49 - 0.44) - 0}{\sqrt{\frac{0.96^2}{20} + \frac{1.4^2}{20}}} = 0.132 \text{ (df = 19)}$$

b. 90% CI: $-0.59 < \mu_1 - \mu_2 < 0.69$ (Table: $-0.61 < \mu_1 - \mu_2 < 0.71$);

$$(\bar{x}_1 - \bar{x}_2) \pm t_{\alpha/2}\sqrt{\frac{s_1^2}{n_1} + \frac{s_2^2}{n_2}} = (0.49 - 0.44) \pm 1.729\sqrt{\frac{0.96^2}{20} + \frac{1.4^2}{20}} \text{ (df = 19)}$$

c. Magnets do not appear to be effective in treating back pain. It is valid to argue that the magnets *might* appear to be effective if the sample sizes were larger.

11. a. H_0: $\mu_1 = \mu_2$; H_1: $\mu_1 \neq \mu_2$; population$_1$ = female, population$_2$ = male;
Test statistic: $t = 0.674$; P-value = 0.5015 (Table: P-value > 0.20); Critical values: $t = \pm 1.979$ (Table: $t = \pm 1.995$); Fail to reject H_0. There is not sufficient evidence to warrant rejection of the claim that females and males have the same mean BMI.

$$t = \frac{(\bar{x}_1 - \bar{x}_2) - (\mu_1 - \mu_2)}{\sqrt{\frac{s_1^2}{n_1} + \frac{s_2^2}{n_2}}} = \frac{(29.10 - 28.38) - 0}{\sqrt{\frac{7.39^2}{70} + \frac{5.37^2}{80}}} = 0.647 \text{ (df = 69)}$$

b. 95% CI: $-1.39 < \mu_1 - \mu_2 < 2.83$ (Table: $-1.41 < \mu_1 - \mu_2 < 2.85$); Because the confidence interval includes 0, there is not sufficient evidence to warrant rejection of the claim that the two samples are from populations with the same mean.

$$(\bar{x}_1 - \bar{x}_2) \pm t_{\alpha/2}\sqrt{\frac{s_1^2}{n_1} + \frac{s_2^2}{n_2}} = (29.10 - 28.38) \pm 1.995\sqrt{\frac{7.39^2}{70} + \frac{5.37^2}{80}} \text{ (df = 69)}$$

c. Based on the available sample data, it appears that males and females have the same mean BMI, but we can only conclude that there isn't sufficient evidence to say that they are different.

13. a. H_0: $\mu_1 = \mu_2$; H_1: $\mu_1 \neq \mu_2$; population$_1$ = female, population$_2$ = male;
Test statistic: $t = -2.025$; P-value = 0.0460 (Table: P-value < 0.05); Critical values: $t = \pm 1.988$ (Table: $t = \pm 2.023$); Reject H_0. There is sufficient evidence to warrant rejection of the claim that the two samples are from populations with the same mean.

$$t = \frac{(\bar{x}_1 - \bar{x}_2) - (\mu_1 - \mu_2)}{\sqrt{\frac{s_1^2}{n_1} + \frac{s_2^2}{n_2}}} = \frac{(3.79 - 4.01) - 0}{\sqrt{\frac{0.51^2}{40} + \frac{0.53^2}{53}}} = 0.647 \text{ (df = 39)}$$

b. 95% CI: $-0.44 < \mu_1 - \mu_2 < 0.00$; Because the confidence interval includes negative numbers only and does not include 0, there is sufficient evidence to warrant rejection of the claim that the two samples are from populations with the same mean.

$$(\bar{x}_1 - \bar{x}_2) \pm t_{\alpha/2}\sqrt{\frac{s_1^2}{n_1} + \frac{s_2^2}{n_2}} = (3.79 - 4.01) \pm 2.023\sqrt{\frac{0.51^2}{40} + \frac{0.53^2}{53}} \text{ (df = 39)}$$

c. Yes, with the smaller samples of size 12 and 15, there was not sufficient evidence to warrant rejection of the null hypothesis, but there is sufficient evidence with the larger samples.

15. a. H_0: $\mu_1 = \mu_2$; H_1: $\mu_1 > \mu_2$; population$_1$ = pre-1964, population$_2$ = post-1964;
Test statistic: $t = 32.771$; P-value = 0.0001 (Table: P-value < 0.005); Critical value: $t = 1.667$ (Table: $t = 1.685$); Reject H_0. There is sufficient evidence to support the claim that pre-1964 quarters have a mean weight that is greater than the mean weight of post-1964 quarters.

$$t = \frac{(\bar{x}_1 - \bar{x}_2) - (\mu_1 - \mu_2)}{\sqrt{\frac{s_1^2}{n_1} + \frac{s_2^2}{n_2}}} = \frac{(6.19267 - 5.63930) - 0}{\sqrt{\frac{0.08700^2}{40} + \frac{0.06194^2}{40}}} = 32.771 \text{ (df = 39)}$$

b. 90% CI: 0.52522 lb < $\mu_1 - \mu_2$ < 0.58125 lb (Table: 0.52492 lb < $\mu_1 - \mu_2$ < 0.58182 lb);

$$(\bar{x}_1 - \bar{x}_2) \pm t_{\alpha/2}\sqrt{\frac{s_1^2}{n_1} + \frac{s_2^2}{n_2}} = (6.19267 - 5.63930) \pm 1.685\sqrt{\frac{0.08700^2}{40} + \frac{0.06194^2}{40}} \text{ (df = 39)}$$

c. Yes, vending machines are not affected very much because pre-1964 quarters are mostly out of circulation.

17. H_0: $\mu_1 = \mu_2$; H_1: $\mu_1 \neq \mu_2$; population$_1$ = female, population$_2$ = male;
Test statistic: $t = -0.315$; P-value = 0.7576 (Table: P-value > 0.20); Critical values ($\alpha = 0.05$): $t = \pm 2.159$ (Table: $t = \pm 2.262$); Fail to reject H_0. There is not sufficient evidence to warrant rejection of the claim that female professors and male professors have the same mean evaluation ratings. There does not appear to be a difference between male and female professor evaluation scores.

$$t = \frac{(\bar{x}_1 - \bar{x}_2) - (\mu_1 - \mu_2)}{\sqrt{\frac{s_1^2}{n_1} + \frac{s_2^2}{n_2}}} = \frac{(4.02 - 4.10) - 0}{\sqrt{\frac{0.7208^2}{10} + \frac{0.3528^2}{10}}} = -0.315 \text{ (df = 9)}$$

19. H_0: $\mu_1 = \mu_2$; H_1: $\mu_1 \neq \mu_2$; population$_1$ = recent, population$_2$ = past;
Test statistic: $t = -2.385$; P-value = 0.0244 (Table: P-value < 0.05); Critical values ($\alpha = 0.05$): $t = \pm 2.052$ (Table: $t = \pm 2.201$); The conclusion depends on the choice of the significance level. There is a significant difference between the two population means at the 0.05 significance level, but not at the 0.01 significance level.

$$t = \frac{(\bar{x}_1 - \bar{x}_2) - (\mu_1 - \mu_2)}{\sqrt{\frac{s_1^2}{n_1} + \frac{s_2^2}{n_2}}} = \frac{(78.82 - 89.08) - 0}{\sqrt{\frac{13.965^2}{17} + \frac{9.190^2}{12}}} = -2.385 \text{ (df = 11)}$$

21. H_0: $\mu_1 = \mu_2$; H_1: $\mu_1 < \mu_2$; population$_1$ = men, population$_2$ = women;
Test statistic: $t = -0.132$; P-value = 0.4477 (Table: P-value > 0.10); Critical value: $t = -1.669$ (Table: $t = -1.688$); Fail to reject H_0. There is not sufficient evidence to support the claim that men talk less than women.

$$t = \frac{(\bar{x}_1 - \bar{x}_2) - (\mu_1 - \mu_2)}{\sqrt{\frac{s_1^2}{n_1} + \frac{s_2^2}{n_2}}} = \frac{(14{,}060.38 - 14{,}296.69) - 0}{\sqrt{\frac{9065.03^2}{37} + \frac{6440.97^2}{42}}} = -0.132 \text{ (df = 36)}$$

23. H_0: $\mu_1 = \mu_2$; H_1: $\mu_1 < \mu_2$; population$_1$ = girls, population$_2$ = boys;
Test statistic: $t = -3.450$; P-value = 0.0003 (Table: P-value < 0.005); Critical value ($\alpha = 0.05$): $t = -1.649$; (Table: $t = -1.653$); Reject H_0. There is sufficient evidence to support the claim that at birth, girls have a lower mean weight than boys.

$$t = \frac{(\bar{x}_1 - \bar{x}_2) - (\mu_1 - \mu_2)}{\sqrt{\frac{s_1^2}{n_1} + \frac{s_2^2}{n_2}}} = \frac{(3037.07 - 3272.82) - 0}{\sqrt{\frac{706.268^2}{205} + \frac{660.154^2}{195}}} = -3.450 \text{ (df = 194)}$$

25. a. H_0: $\mu_1 = \mu_2$; H_1: $\mu_1 > \mu_2$; population$_1$ = low lead, population$_2$ = high lead;
Test statistic: $t = 1.705$; P-value = 0.0457 (Table: P-value < 0.05); Critical value: $t = 1.661$ (Table: $t = 1.987$); df $= 78 + 21 - 2 = 977$; Reject H_0. There is sufficient evidence to support the claim that the mean IQ score of people with low blood lead levels is higher than the mean IQ score of people with high blood lead levels.

$$s_p^2 = \frac{(n_1 - 1)s_1^2 + (n_2 - 1)s_2^2}{(n_1 - 1) + (n_2 - 1)} = \frac{(78-1)15.34451^2 + (21-1)8.988352^2}{(78-1) + (21-1)} = 203.564;$$

$$t = \frac{(\bar{x}_1 - \bar{x}_2) - (\mu_1 - \mu_2)}{\sqrt{\frac{s_p^2}{n_1} + \frac{s_p^2}{n_2}}} = \frac{(92.88462 - 86.90476) - 0}{\sqrt{\frac{203.564}{78} + \frac{203.564}{21}}} = 1.705 \text{ (df = 97)}$$

b. 90% CI: $0.15 < \mu_1 - \mu_2 < 11.81$;

$$(\bar{x}_1 - \bar{x}_2) \pm t_{\alpha/2}\sqrt{\frac{s_1^2}{n_1} + \frac{s_2^2}{n_2}} = (92.88462 - 86.90476) \pm 1.661\sqrt{\frac{203.564}{78} + \frac{203.564}{21}} \text{ (df = 97)}$$

c. Yes, it does appear that exposure to lead has an effect on IQ scores.
With pooling, df increases dramatically to 97, but the test statistic decreases from 2.282 to 1.705 (because the estimated standard deviation increases from 2.620268 to 3.507614), the P-value increases to 0.0457, and the 90% confidence interval becomes wider. With pooling, these results do not show greater significance.

27. H_0: $\mu_1 = \mu_2$; H_1: $\mu_1 \neq \mu_2$; population$_1$ = treatment, population$_2$ = placebo;
Test statistic: $t = 15.322$; P-value = 0.0000 (Table: P-value < 0.01); Critical values: $t = \pm 2.080$; Reject H_0. There is sufficient evidence to warrant rejection of the claim that the two populations have the same mean.

$$s_p^2 = \frac{(n_1 - 1)s_1^2 + (n_2 - 1)s_2^2}{(n_1 - 1) + (n_2 - 1)} = \frac{(22-1)0.015^2 + (22-1)0^2}{(22-1) + (22-1)} = 0.000125;$$

$$t = \frac{(\bar{x}_1 - \bar{x}_2) - (\mu_1 - \mu_2)}{\sqrt{\frac{s_p^2}{n_1} + \frac{s_p^2}{n_2}}} = \frac{(0.049 - 0.000) - 0}{\sqrt{\frac{0.000125}{22} + \frac{0.000125}{22}}} = 15.322$$

Section 9-3: Two Dependent Samples (Matched Pairs)

1. Only parts (a) and (c) are true.
3. The results will be the same.
5. a. H_0: $\mu_d = 0$ year; H_1: $\mu_d < 0$ year; difference = actress − actor;
Test statistic: $t = -2.609$; P-value = 0.0142 (Table: P-value < 0.025); Critical value: $t = -1.833$;
Reject H_0. There is sufficient evidence to support the claim that for the population of ages of Best Actresses and Best Actors, the differences have a mean less than 0. There is sufficient evidence to conclude that Best Actresses are generally younger than Best Actors.
$$t = \frac{\bar{d} - \mu_d}{s_d/\sqrt{n}} = \frac{-9.7 - 0}{11.757/\sqrt{10}} = -2.609 \text{ (df = 9)}$$
b. 90% CI: $-16.5 \text{ years} < \mu_d < -2.9 \text{ years}$; The confidence interval consists of negative numbers only and does not include 0.
$$\bar{d} \pm t_{\alpha/2}\frac{s_d}{\sqrt{n}} = -9.7 \pm 2.262\frac{11.757}{\sqrt{10}} \text{ (df = 9)}$$
7. a. H_0: $\mu_d = 0°\text{F}$; H_1: $\mu_d \neq 0°\text{F}$; difference = 8 AM − 12 AM;
Test statistic: $t = -7.499$; P-value = 0.0003 (Table: P-value < 0.01); Critical values: $t = \pm 2.447$; Reject H_0. There is sufficient evidence to warrant rejection of the claim that there is no difference between body temperatures measured at 8 AM and at 12 AM. There appears to be a difference.
$$t = \frac{\bar{d} - \mu_d}{s_d/\sqrt{n}} = \frac{-1.49 - 0}{0.524/\sqrt{7}} = -7.499 \text{ (df = 6)}$$
b. 95% CI: $-1.97°\text{F} < \mu_d < -1.00°\text{F}$; The confidence interval consists of negative numbers only and does not include 0.
$$\bar{d} \pm t_{\alpha/2}\frac{s_d}{\sqrt{n}} = -1.49 \pm 2.447\frac{0.524}{\sqrt{7}} \text{ (df = 6)}$$
9. H_0: $\mu_d = 0$ in.; H_1: $\mu_d \neq 0$ in.; difference = mother − daughter;
Test statistic: $t = -7.499$; P-value = 0.2013 (Table: P-value > 0.20); Critical values: $t = \pm 2.262$; Fail to reject H_0. There is not sufficient evidence to warrant rejection of the claim that there is no difference in heights between mothers and their first daughters.
$$t = \frac{\bar{d} - \mu_d}{s_d/\sqrt{n}} = \frac{-0.95 - 0}{2.179/\sqrt{10}} = -1.379 \text{ (df = 9)}$$
11. H_0: $\mu_d = 0$; H_1: $\mu_d \neq 0$; difference = male by female − female by male;
Test statistic: $t = 0.793$; P-value = 0.4509 (Table: P-value > 0.20); Critical values: $t = \pm 2.306$; Fail to reject H_0. There is not sufficient evidence to support the claim that there is a difference between female attribute ratings and male attribute ratings.
$$t = \frac{\bar{d} - \mu_d}{s_d/\sqrt{n}} = \frac{1.89 - 0}{7.149/\sqrt{9}} = 0.793 \text{ (df = 8)}$$
13. 95% CI: $-6.5 \text{ admissions} < \mu_d < -0.2 \text{ admissions}$; Because the confidence interval does not include 0 admission, it appears that there is sufficient evidence to warrant rejection of the claim that when the 13th day of a month falls on a Friday, the numbers of hospital admissions from motor vehicle crashes are not affected. Hospital admissions do appear to be affected.
$$\bar{d} \pm t_{\alpha/2}\frac{s_d}{\sqrt{n}} = -3.333 \pm 2.571\frac{3.011}{\sqrt{6}} \text{ (difference = 6th − 13th, df = 5)}$$

15. 95% CI: $0.69 < \mu_d < 5.66$; Because the confidence interval limits do not contain 0 and they consist of positive values only, it appears that the "before" measurements are greater than the "after" measurements, so hypnotism does appear to be effective in reducing pain.

$$\bar{d} \pm t_{\alpha/2} \frac{s_d}{\sqrt{n}} = 3.125 \pm 2.365 \frac{2.911}{\sqrt{8}} \text{ (difference = before − after, df = 7)}$$

17. a. H_0: $\mu_d = 0$ year; H_1: $\mu_d < 0$ year; difference = actress − actor;
Test statistic: $t = -5.185$; P-value = 0.0000 (Table: P-value < 0.005); Critical value: $t = -1.663$ (Table: $t = -1.662$); Reject H_0. There is sufficient evidence to support the claim that actresses are generally younger than actors.

$$t = \frac{\bar{d} - \mu_d}{s_d/\sqrt{n}} = \frac{-7.89 - 0}{14.18/\sqrt{87}} = -5.185 \text{ (df = 86)}$$

b. 90% CI: -10.4 years $< \mu_d < -5.4$ years; The confidence interval consists of negative numbers only and does not include 0.

$$\bar{d} \pm t_{\alpha/2} \frac{s_d}{\sqrt{n}} = -7.89 \pm 1.663 \frac{14.18}{\sqrt{87}} \text{ (df = 86)}$$

19. a. H_0: $\mu_d = 0°\text{F}$; H_1: $\mu_d \neq 0°\text{F}$; difference = 8 AM − 12 AM;
Test statistic: $t = -8.485$; P-value = 0.0000 (Table: P-value < 0.01); Critical values: $t = \pm 1.996$ (Table: $t = \pm 1.994$); Reject H_0. There is sufficient evidence to warrant rejection of the claim of no difference between body temperatures measured at 8 am and at 12 am. There appears to be a difference.

$$t = \frac{\bar{d} - \mu_d}{s_d/\sqrt{n}} = \frac{-0.85 + 0}{0.833/\sqrt{69}} = -8.485 \text{ (df = 68)}$$

b. 95% CI: $-1.05°\text{F} < \mu_d < -0.65°\text{F}$; The confidence interval consists of negative numbers only and does not include 0.

$$\bar{d} \pm t_{\alpha/2} \frac{s_d}{\sqrt{n}} = -0.85 \pm 1.994 \frac{0.833}{\sqrt{69}} \text{ (df = 68)}$$

21. H_0: $\mu_d = 0$ in.; H_1: $\mu_d \neq 0$ in.; difference = mother − daughter;
Test statistic: $t = -4.090$; P-value = 0.0001 (Table: P-value < 0.01); Critical values: $t = \pm 1.978$ (Table: $t \approx \pm 1.974$); Reject H_0. There is sufficient evidence to warrant rejection of the claim of no difference in heights between mothers and their first daughters.

$$t = \frac{\bar{d} - \mu_d}{s_d/\sqrt{n}} = \frac{-0.93 - 0}{2.636/\sqrt{134}} = -4.090 \text{ (df = 133)}$$

23. H_0: $\mu_d = 0$; H_1: $\mu_d \neq 0$; difference = male by female − female by male;
Test statistic: $t = 0.191$; P-value = 0.8485 (Table: P-value > 0.20); Critical values: $t = \pm 1.972$; Fail to reject H_0. There is not sufficient evidence to support the claim that there is a difference between female attribute ratings and male attribute ratings.

$$t = \frac{\bar{d} - \mu_d}{s_d/\sqrt{n}} = \frac{0.111 - 0}{8.154/\sqrt{199}} = 0.793 \text{ (df = 198)}$$

25. For the temperatures in degrees Fahrenheit and the temperatures in degrees Celsius, the test statistic of $t = 0.124$ is the same, the P-value of 0.9023 is the same, the critical values of $t = \pm 2.028$ are the same, and the conclusions are the same, so the hypothesis test results are the same in both cases. The confidence intervals are $-0.25°\text{F} < \mu_d < 0.28°\text{F}$ and $-0.14°\text{C} < \mu_d < 0.16°\text{C}$. The confidence interval limits of $-0.14°\text{C}$ and $0.16°\text{C}$ have numerical values that are $5/9$ of the numerical values of $-0.25°\text{F}$ and $0.28°\text{F}$.

Section 9-4: Two Variances or Standard Deviations

1. a. No, the numerator will always be larger than the denominator in the fraction.
   b. No, both variances are nonnegative, so their quotient cannot be negative.
   c. The two samples have standard deviations (or variances) that are very close in value.
   d. skewed right

3. No, unlike some other tests that have a requirement that samples must be from normally distributed populations or the samples must have more than 30 values, the F test has a requirement that the samples must be from normally distributed populations, regardless of how large the samples are.

5. H_0: $\sigma_1 = \sigma_2$; H_1: $\sigma_1 \neq \sigma_2$; population$_1$ = red, population$_2$ = blue;
   Test statistic: $F = s_1^2 / s_2^2 = 0.97^2 / 0.63^2 = 2.3706$; P-value = 0.0129; Upper critical value: $F = 1.9678$ (Table: $1.8752 < F < 2.0739$); Reject H_0. There is sufficient evidence to warrant rejection of the claim that creative task scores have the same variation with a red background and a blue background.

7. H_0: $\sigma_1 = \sigma_2$; H_1: $\sigma_1 > \sigma_2$; population$_1$ = treatment, population$_2$ = placebo;
   Test statistic: $F = s_1^2 / s_2^2 = 2.20^2 / 0.72^2 = 9.3364$; P-value = 0.0000; Critical value: $F = 2.0842$ (Table: $2.0540 < F < 2.0960$); Reject H_0. There is sufficient evidence to support the claim that the treatment group has errors that vary more than the errors of the placebo group.

9. H_0: $\sigma_1 = \sigma_2$; H_1: $\sigma_1 \neq \sigma_2$; population$_1$ = regular Coke, population$_2$ = Diet Coke;
   Test statistic: $F = s_1^2 / s_2^2 = 0.00751^2 / 0.00439^2 = 2.9265$; P-value = 0.0020; Upper critical value: $F = 1.9611$ (Table: $1.8752 < F < 2.0739$); Reject H_0. There is sufficient evidence to warrant rejection of the claim that variation is the same for both types of Coke.

11. H_0: $\sigma_1 = \sigma_2$; H_1: $\sigma_1 > \sigma_2$; population$_1$ = sham treatment, population$_2$ = magnet treatment;
    Test statistic: $F = s_1^2 / s_2^2 = 1.4^2 / 0.96^2 = 2.1267$; P-value = 0.0543; Critical value: $F = 2.1682$ (Table: $2.1555 < F < 2.2341$); Fail to reject H_0. There is not sufficient evidence to support the claim that those given a sham treatment have pain reductions that vary more than the pain reductions for those treated with magnets.

13. H_0: $\sigma_1 = \sigma_2$; H_1: $\sigma_1 \neq \sigma_2$; population$_1$ = female, population$_2$ = male;
    Test statistic: $F = s_1^2 / s_2^2 = 0.721^2 / 0.353^2 = 4.1750$; P-value = 0.0447; Upper critical value: $F = 4.0260$; Reject H_0. There is sufficient evidence to warrant rejection of the claim that female professors and male professors have evaluation scores with the same variation.

15. H_0: $\sigma_1 = \sigma_2$; H_1: $\sigma_1 \neq \sigma_2$; population$_1$ = recent, population$_2$ = past;
    Test statistic: $F = s_1^2 / s_2^2 = 13.965^2 / 9.190^2 = 2.3095$; P-value = 0.1635; Upper critical value $(\alpha = 0.05)$: $F = 3.3044$ (Table: $3.2261 < F < 3.3299$); Fail to reject H_0. There is not sufficient evidence to support a claim that the variation of the times between eruptions has changed.

17. a. Calculations not shown.
    b. $c_1 = 3, c_2 = 0$
    c. Critical value $= \dfrac{\log(0.05/2)}{\log\left(\dfrac{25}{25+16}\right)} = 7.4569$
    d. $c_1 = 3 < 7.4569$; Fail to reject H_0. There is not sufficient evidence to support a claim that the two populations of scores have different amounts of variation.

19. $F_L = \dfrac{1}{2.4374} = 0.4103, F_R = 2.7006$

Quick Quiz

1. $H_0: p_1 = p_2$; $H_1: p_1 \neq p_2$; population$_1$ = women, population$_2$ = men;
2. $x_1 = 258, x_2 = 282, \hat{p}_1 = \frac{258}{1121} = 0.230, \hat{p}_2 = \frac{282}{1084} = 0.260, \overline{p} = \frac{258+282}{1121+1084} = 0.245$
3. P-value = 0.1015 (Table: 0.1010)
4. a. 95% CI: $-0.0659 < p_1 - p_2 < 0.00591$

   b. The confidence interval includes the value of 0, so it is possible that the two proportions are equal. There is not a significant difference.
5. Fail to reject H_0. There is not sufficient evidence to warrant rejection of the claim that for the people who were aware of the statement, the proportion of women is equal to the proportion of men.
6. True, since $n > 30$.
7. False, the requirements are $np \geq 5$ and $nq \geq 5$.
8. Because the data consist of matched pairs, they are dependent.
9. $H_0: \mu_d = 0$; $H_1: \mu_d \neq 0$; difference = right arm − left arm;
10. a. $t = \frac{\overline{d} - \mu_d}{s_d/\sqrt{n}} = \frac{0.111 - 0}{8.154/\sqrt{199}}$

    b. $t = \frac{(\overline{x}_1 - \overline{x}_2) - (\mu_1 - \mu_2)}{\sqrt{\frac{s_1^2}{n_1} + \frac{s_2^2}{n_2}}}$

    c. $z = \frac{(\hat{p}_1 - \hat{p}_2) - (p_1 - p_2)}{\sqrt{\frac{\overline{p}\overline{q}}{n_1} + \frac{\overline{p}\overline{q}}{n_2}}}$

    d. $F = \frac{s_1^2}{s_2^2}$

Review Exercises

1. $H_0: p_1 = p_2$; $H_1: p_1 < p_2$; population$_1$ = \$1 bill, population$_2$ = 4 quarters;
   Test statistic: $z = -3.49$; P-value = 0.0002; Critical value: $z = -1.645$; Reject H_0. There is sufficient evidence to support the claim that money in a large denomination is less likely to be spent relative to an equivalent amount in smaller denominations.

$$\overline{p} = \frac{12+27}{46+43} = \frac{39}{89}; \overline{q} = 1 - \frac{39}{89} = \frac{50}{89};$$

$$z = \frac{(\hat{p}_1 - \hat{p}_2) - (p_1 - p_2)}{\sqrt{\frac{\overline{p}\overline{q}}{n_1} + \frac{\overline{p}\overline{q}}{n_2}}} = \frac{\left(\frac{12}{46} - \frac{27}{43}\right) - 0}{\sqrt{\frac{\left(\frac{39}{89}\right)\left(\frac{50}{89}\right)}{46} + \frac{\left(\frac{39}{89}\right)\left(\frac{50}{89}\right)}{43}}} = -3.49$$

2. 90% CI: $-0.528 < p_1 - p_2 < -0.206$; The confidence interval limits do not contain 0, so it appears that there is a significant difference between the two proportions. Because the confidence interval consists of negative values only, it appears that p1 is less than p2, so it appears that money in a large denomination is less likely to be spent relative to an equivalent amount in smaller denominations

$$(\hat{p}_1 - \hat{p}_2) \pm z_{\alpha/2}\sqrt{\frac{\hat{p}_1\hat{q}_1}{n_1} + \frac{\hat{p}_2\hat{q}_2}{n_2}} = \left(\frac{12}{46} - \frac{27}{43}\right) \pm 1.645\sqrt{\frac{\left(\frac{12}{46}\right)\left(\frac{34}{46}\right)}{46} + \frac{\left(\frac{27}{43}\right)\left(\frac{16}{43}\right)}{43}}$$

3. 95% CI: $-25.33 \text{ cm} < \mu_1 - \mu_2 < -7.51 \text{ cm}$ (Table: $-25.70 \text{ cm} < \mu_1 - \mu_2 < -7.14 \text{ cm}$); With 95% confidence, we conclude that the mean height of women is less than the mean height of men by an amount that is between 7.51 cm and 25.33 cm (Table: 7.14 cm and 25.70 cm).

$$(\bar{x}_1 - \bar{x}_2) \pm t_{\alpha/2}\sqrt{\frac{s_1^2}{n_1} + \frac{s_2^2}{n_2}} = (162.35 - 178.77) \pm 2.262\sqrt{\frac{11.847^2}{10} + \frac{5.302^2}{10}} \text{ (df} = 9);$$

population$_1$ = women, population$_2$ = men

4. H_0: $\mu_1 = \mu_2$; H_1: $\mu_1 < \mu_2$; population$_1$ = women, population$_2$ = men;
Test statistic: $t = -4.001$; P-value = 0.0008 (Table: P-value < 0.005); Critical value: $t = -2.666$ (Table: $t = -2.821$); Reject H_0. There is sufficient evidence to support the claim that women have heights with a mean that is less than the mean height of men.

$$t = \frac{(\bar{x}_1 - \bar{x}_2) - (\mu_1 - \mu_2)}{\sqrt{\frac{s_1^2}{n_1} + \frac{s_2^2}{n_2}}} = \frac{(162.35 - 178.77) - 0}{\sqrt{\frac{11.847^2}{10} + \frac{5.302^2}{10}}} = -4.001 \text{ (df} = 9)$$

5. H_0: $\mu_d = 0$ cm; H_1: $\mu_d > 0$ cm; difference = before − after;
Test statistic: $t = 6.371$; P-value = 0.0000 (Table: P-value < 0.005); Critical value: $t = 2.718$; Reject H_0. There is sufficient evidence to support the claim that captopril is effective in lowering systolic blood pressure.

$$t = \frac{\bar{d} - \mu_d}{s_d/\sqrt{n}} = \frac{18.583 - 0}{10.104/\sqrt{12}} = 6.371 \text{ (df} = 11)$$

6. H_0: $\mu_1 = \mu_2$; H_1: $\mu_1 > \mu_2$; population$_1$ = nonstress, population$_2$ = stress;
Test statistic: $t = 2.879$; P-value = 0.0026 (Table: P-value < 0.005); Critical value: $t = 2.376$ (Table: $t = 2.429$); Reject H_0. There is sufficient evidence to support the claim that the mean number of details recalled is lower for the stress group. It appears that "stress decreases the amount recalled," but we should not conclude that stress is the cause of the decrease.

$$t = \frac{(\bar{x}_1 - \bar{x}_2) - (\mu_1 - \mu_2)}{\sqrt{\frac{s_1^2}{n_1} + \frac{s_2^2}{n_2}}} = \frac{(53.3 - 45.3) - 0}{\sqrt{\frac{11.6^2}{40} + \frac{13.2^2}{40}}} = 2.429 \text{ (df} = 39)$$

7. H_0: $\mu_d = 0$ cm; H_1: $\mu_d > 0$ cm; difference = 1 day − 30 days;
Test statistic: $t = 14.061$; P-value = 0.0000 (Table: P-value < 0.005); Critical value: $t = 3.365$; Reject H_0. There is sufficient evidence to support the claim that flights scheduled 1 day in advance cost more than flights scheduled 30 days in advance. Save money by scheduling flights 30 days in advance.

$$t = \frac{\bar{d} - \mu_d}{s_d/\sqrt{n}} = \frac{419.17 - 0}{73.022/\sqrt{6}} = 14.061 \text{ (df} = 5)$$

8. H_0: $\sigma_1 = \sigma_2$; H_1: $\sigma_1 \neq \sigma_2$; population$_1$ = women, population$_2$ = men;
Test statistic: $F = s_1^2/s_2^2 = 11.847^2/5.302^2 = 4.9933$; P-value = 0.0252; Upper critical value : $F = 4.0260$; Reject H_0. There is sufficient evidence to warrant rejection of the claim that women and men have heights with the same variation.

Cumulative Review Exercises

1. a. Because the sample data are matched with each column consisting of heights from the same family, the data are dependent.

   b. $\bar{x} = \dfrac{64.0+68.0+70.0+71.0+71.0+71.0+71.0+71.7}{8} = 69.7$ in.; $Q_2 = \dfrac{71.0+71.0}{2} = 71.0$ in.;

   range $= 71.7 - 64.0 = 7.7$ in.; $s = \sqrt{\dfrac{(64.0-69.7)^2 + \cdots + (71.7-69.7)^2}{14-1}} = 2.6$ in.; $s^2 = 6.6\text{ in}^2$

   c. ratio

   d. continuous

2. There does not appear to be a correlation or association between the heights of fathers and the heights of their sons.

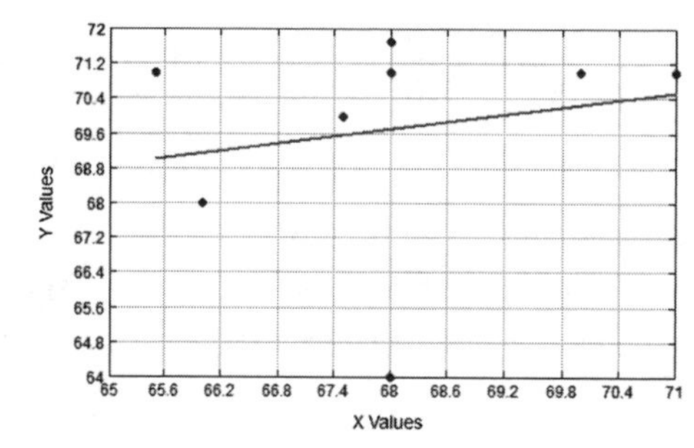

3. 90% CI: $\bar{x} \pm t_{\alpha/2}\dfrac{s}{\sqrt{n}} = 69.7 \pm 2.306 \cdot \dfrac{2.570}{\sqrt{8}} \Rightarrow 67.6 \text{ in.} < \mu < 71.9 \text{ in.}$; We have 95% confidence that the limits of 67.6 in. and 71.9 in. actually contain the true value of the mean height of all adult sons.

4. $H_0\colon \mu_d = 0$ in.; $H_1\colon \mu_d \neq 0$ in.; difference = father − son;
   Test statistic: $t = -1.712$; P-value $= 0.1326$ (Table: P-value > 0.10); Critical values: $t = \pm 2.365$; Fail to reject H_0. There is not sufficient evidence to warrant rejection of the claim that differences between heights of fathers and their sons have a mean of 0. There does not appear to be a difference between heights of fathers and their sons.

   $$t = \frac{\bar{d} - \mu_d}{s_d/\sqrt{n}} = \frac{-1.7125 - 0}{2.847/\sqrt{8}} = -1.702 \text{ (df} = 7)$$

5. Because the points lie reasonably close to a straight-line pattern, and there is no other pattern that is not a straight-line pattern and there are no outliers, the sample data appear to be from a population with a normal distribution.

6. The shape of the histogram indicates that the sample data appear to be from a population with a distribution that is approximately normal.

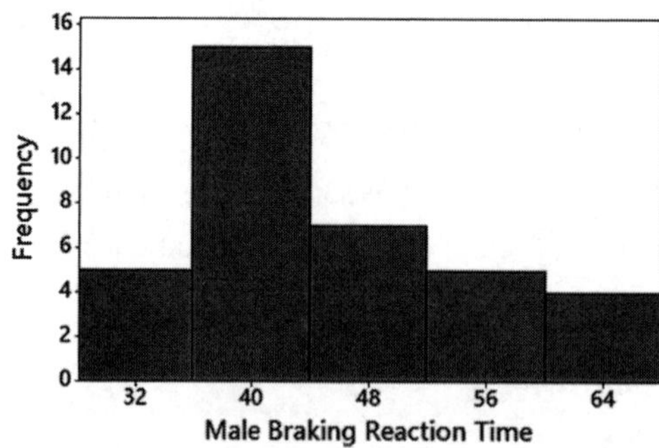

7. Because the points are reasonably close to a straight-line pattern and there is no other pattern that is not a straight-line pattern, it appears that the braking reaction times of females are from a population with a normal distribution.

8. Because the boxplots overlap, there does not appear to be a significant difference between braking reaction times of males and females, but the braking reaction times for males appear to be generally lower than the braking reaction times of females.

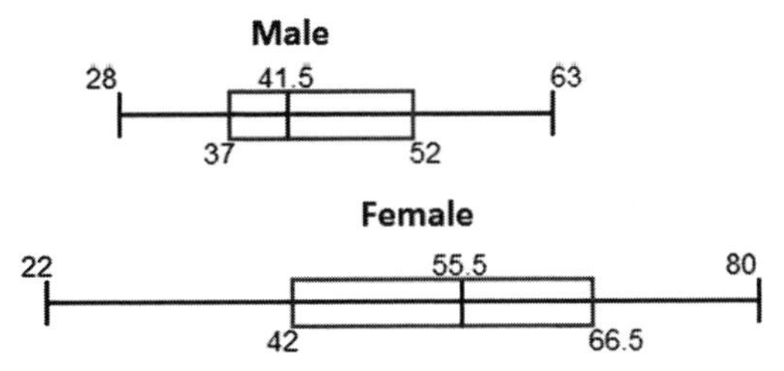

9. H_0: $\mu_1 = \mu_2$; H_1: $\mu_1 \neq \mu_2$; population$_1$ = male, population$_2$ = female;
Test statistic: $t = -3.259$; P-value = 0.0019 (Table: P-value < 0.005); Critical values $t = \pm 2.664$ (Table: $t = \pm 2.724$); Reject H_0. There is sufficient evidence to warrant rejection of the claim that males and females have the same mean braking reaction time. Males appear to have lower reaction times.

$$t = \frac{(\bar{x}_1 - \bar{x}_2) - (\mu_1 - \mu_2)}{\sqrt{\frac{s_1^2}{n_1} + \frac{s_2^2}{n_2}}} = \frac{(44.361 - 54.278) - 0}{\sqrt{\frac{9.472^2}{36} + \frac{15.611^2}{36}}} = -3.259 \text{ (df = 35)}$$

10. a. The sample sizes are greater than 30 and the data appear to be from a populations that meet the loose requirement of being normally distributed.

$$\text{Males: 99\%\% CI: } \bar{x} \pm t_{\alpha/2} \frac{s}{\sqrt{n}} = 44.631 \pm 2.724 \cdot \frac{9.472}{\sqrt{36}} \Rightarrow 40.1 < \mu < 48.7$$

$$\text{Females: 99\%\% CI: } \bar{x} \pm t_{\alpha/2} \frac{s}{\sqrt{n}} = 54.278 \pm 2.724 \cdot \frac{15.611}{\sqrt{36}} \Rightarrow 47.2 < \mu < 61.4$$

The confidence intervals overlap, so there does not appear to be significant difference between the mean braking reaction times of males and females.

b. 99% CI: $-18.0 < \mu_1 - \mu_2 < -1.8$ (Table: $-18.2 < \mu_1 - \mu_2 < -1.6$); Because the confidence interval consists of negative numbers and does not include 0, there appears to be a significant difference between the mean braking reaction times of males and females.

$$(\bar{x}_1 - \bar{x}_2) \pm t_{\alpha/2} \sqrt{\frac{s_1^2}{n_1} + \frac{s_2^2}{n_2}} = (44.361 - 54.278) \pm 2.724 \sqrt{\frac{9.472^2}{36} + \frac{15.611^2}{36}} \text{ (df = 35);}$$

population$_1$ = men, population$_2$ = women

c. The results from part (b) are better.

Chapter 10: Correlation and Regression

Section 10-1: Correlation

1. a. r is a statistic that represents the value of the linear correlation coefficient computed from the paired sample data, and ρ is a parameter that represents the value of the linear correlation coefficient that would be computed by using all of the paired data in the population of all statistics students.
   b. The value of r is estimated to be 0, because it is likely that there is no correlation between body temperature and head circumference.
   c. The value of r does not change if the body temperatures are converted to Fahrenheit degrees.
3. No, a correlation between two variables indicates that they are somehow associated, but that association does not necessarily imply that one of the variables has a direct effect on the other variable. Correlation does not imply causality.
5. $r = 0.963$; P-value $= 0.000$; Critical values: $r = \pm 0.268$ (Table: $r \approx \pm 0.279$); Yes, there is sufficient evidence to support the claim that there is a linear correlation between the weights of bears and their chest sizes. It is easier to measure the chest size of a bear than the weight, which would require lifting the bear onto a scale. It does appear that chest size could be used to predict weight.
7. $r = 0.117$; P-value > 0.05; Critical values: $r = \pm 0.250$ (Table: $r \approx \pm 0.254$); No, there is not sufficient evidence to support the claim that there is a linear correlation between weights of discarded paper and glass.
9. a.

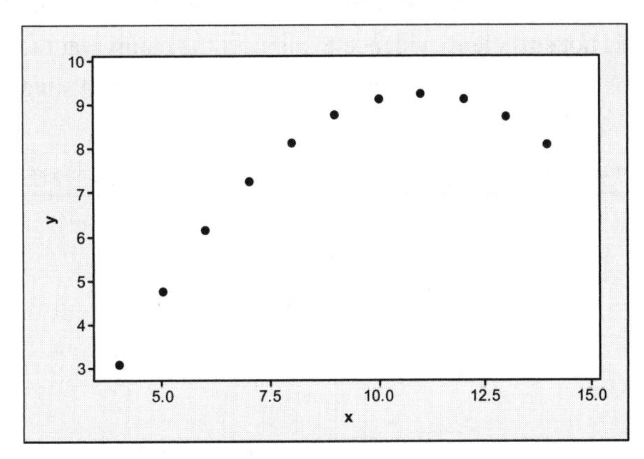

   b. $H_0: \rho = 0$; $H_1: \rho \neq 0$; $r = 0.816$; P-value $= 0.002$ (Table: P-value < 0.01); Critical values ($\alpha = 0.05$): $r = \pm 0.602$; There is sufficient evidence to support the claim of a linear correlation between the two variables.
   c. The scatterplot reveals a distinct pattern that is not a straight line pattern.
11. a. Answer will vary, but because there appears to be an upward pattern, it is reasonable to think that there is a linear correlation.
   b. $H_0: \rho = 0$; $H_1: \rho \neq 0$; $r = 0.906$; Critical values ($\alpha = 0.05$): $r = \pm 0.632$; P-value $= 0.000$ (Table: P-value < 0.01); There is sufficient evidence to support the claim of a linear correlation.
   c. $H_0: \rho = 0$; $H_1: \rho \neq 0$; $r = 0$; Critical values ($\alpha = 0.05$): $r = \pm 0.666$; P-value $= 1.000$ (Table: P-value > 0.05); There is not sufficient evidence to support the claim of a linear correlation.
   d. The effect from a single pair of values can be very substantial, and it can change the conclusion.

13. H_0: $\rho = 0$; H_1: $\rho \neq 0$; $r = 0.799$; P-value $= 0.056$ (Table: P-value > 0.05); Critical values: $r = \pm 0.811$; There is not sufficient evidence to support the claim that there is a linear correlation between Internet users and Nobel Laureates.

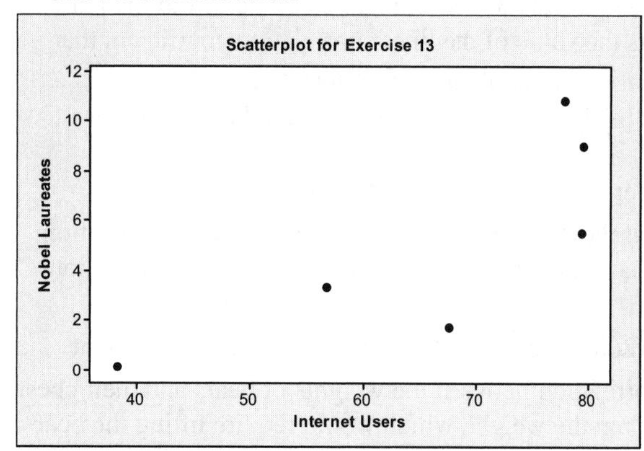

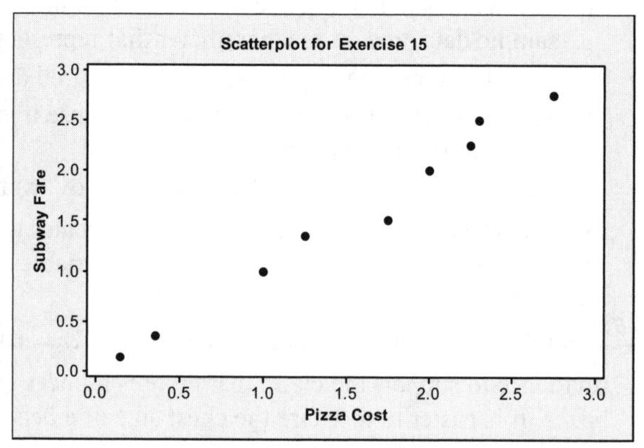

15. H_0: $\rho = 0$; H_1: $\rho \neq 0$; $r = 0.992$; P-value $= 0.000$ (Table: P-value < 0.01); Critical values: $r = \pm 0.666$; Reject H_0. There is sufficient evidence to support the claim that there is a significant linear correlation between the cost of a slice of pizza and the subway fare.

17. H_0: $\rho = 0$; H_1: $\rho \neq 0$; $r = 0.591$; P-value $= 0.294$ (Table: P-value > 0.05); Critical values: $r = \pm 0.878$; Fail to reject H_0. There is not sufficient evidence to support the claim that there is a linear correlation between shoe print lengths and heights of males. The given results do not suggest that police can use a shoe print length to estimate the height of a male.

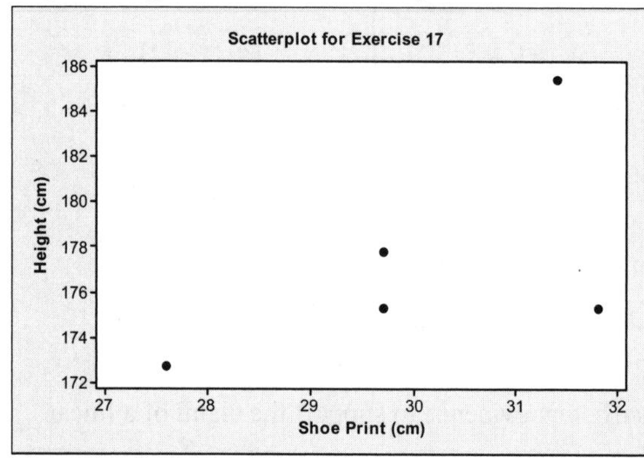

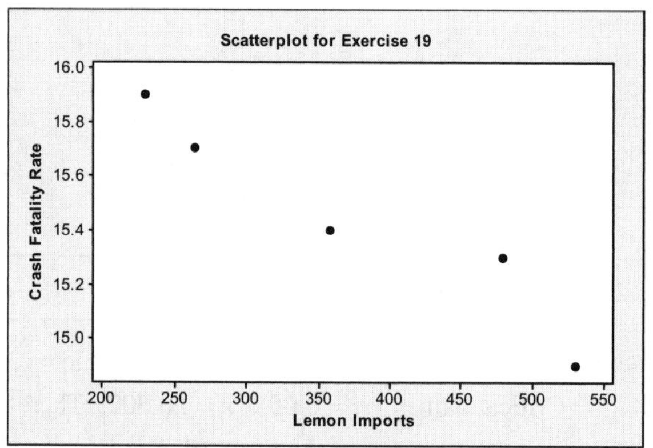

19. H_0: $\rho = 0$; H_1: $\rho \neq 0$; $r = -0.959$; P-value $= 0.010$; Critical values: $r = \pm 0.878$; Reject H_0. There is sufficient evidence to support the claim that there is a linear correlation between weights of lemon imports from Mexico and U.S. car fatality rates. The results do not suggest any cause-effect relationship between the two variables.

21. H_0: $\rho = 0$; H_1: $\rho \neq 0$; $r = -0.288$; P-value $= 0.365$ (Table: P-value > 0.05); Critical values: $r = \pm 0.576$; Fail to reject H_0. There is not sufficient evidence to support the claim that there is a significant linear correlation between the ages of Best Actresses and Best Actors. Because Best Actresses and Best Actors typically appeared in different movies, we should not expect that there would be a correlation between their ages at the time that they won the awards. (See graph on next page.)

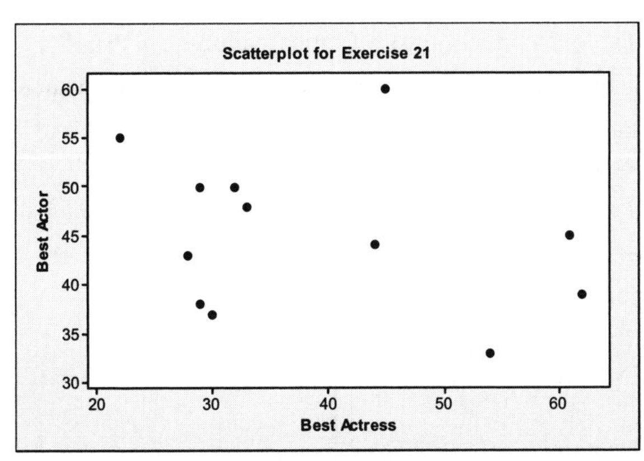

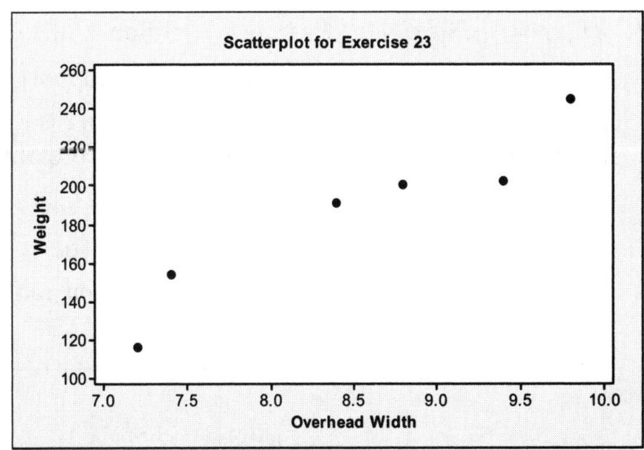

23. H_0: $\rho = 0$; H_1: $\rho \neq 0$; $r = 0.948$; P-value $= 0.004$ (Table: P-value < 0.01); Critical values: $r = \pm 0.811$; Reject H_0. There is sufficient evidence to support the claim of a linear correlation between the overhead width of a seal in a photograph and the weight of a seal.

25. H_0: $\rho = 0$; H_1: $\rho \neq 0$; $r = 0.828$; P-value $= 0.042$ (Table: P-value < 0.05); Critical values: $r = \pm 0.811$; Reject H_0. There is sufficient evidence to support the claim there is a linear correlation between the bill amounts and the tip amounts. If everyone were to tip the same percentage, r would be 1.

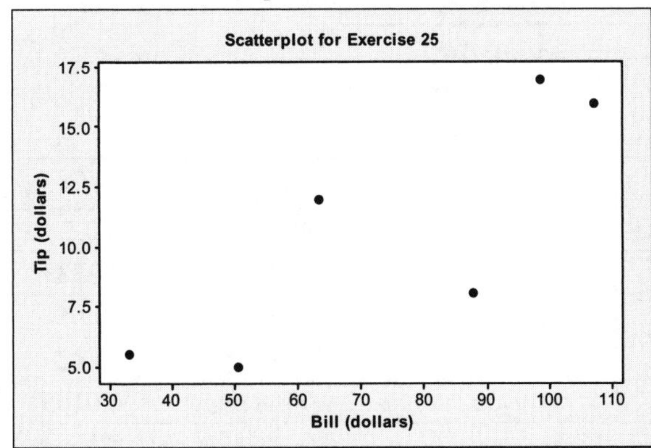

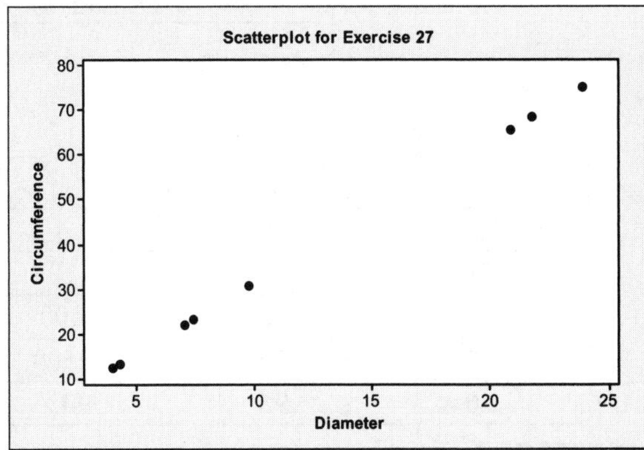

27. H_0: $\rho = 0$; H_1: $\rho \neq 0$; $r = 1.000$; P-value $= 0.000$ (Table: P-value < 0.01); Critical values: $r = \pm 0.707$; Reject H_0. There is sufficient evidence to support the claim that there is a linear correlation between diameters and circumferences. The scatterplot confirms a linear association.

29. H_0: $\rho = 0$; H_1: $\rho \neq 0$; $r = 0.702$; P-value $= 0.000$ (Table: P-value < 0.01); Critical values: $r = \pm 0.413$ (Table: $-0.444 < r < -0.396$ or $0.396 < r < 0.444$); Reject H_0. There is sufficient evidence to support the claim that there is a linear correlation between Internet users and Nobel Laureates.

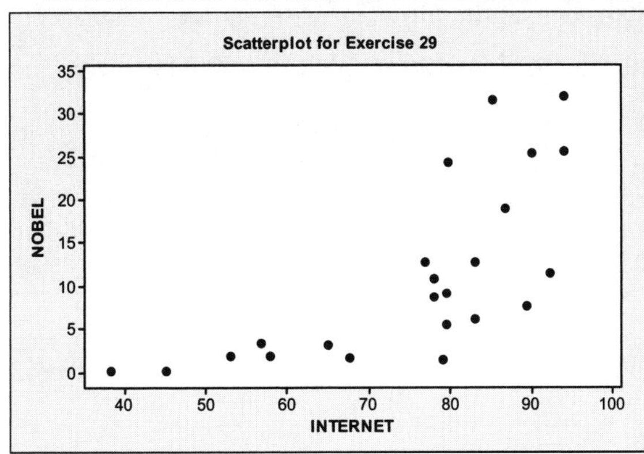

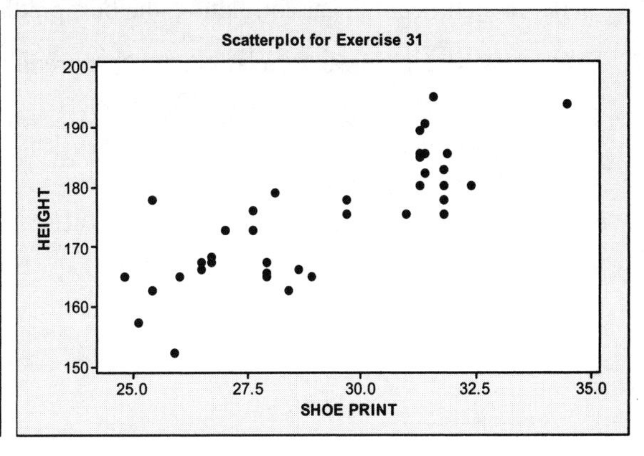

31. H_0: $\rho = 0$; H_1: $\rho \neq 0$; $r = 0.594$; P-value $= 0.007$ (Table: P-value < 0.01); Critical values: $r \approx \pm 0.456$; Reject H_0. There is sufficient evidence to support the claim that there is a linear correlation between shoe print lengths and heights of males. The given results do suggest that police can use a shoe print length to estimate the height of a male. (See graph on previous page.)
33. H_0: $\rho = 0$; H_1: $\rho \neq 0$; $r = 0.319$; P-value $= 0.017$ (Table: P-value < 0.05); Critical values: $r \approx \pm 0.263$ (Table: $r \approx \pm 0.254$); Reject H_0. There is sufficient evidence to support the claim of a linear correlation between the numbers of words spoken by men and women who are in couple relationships.

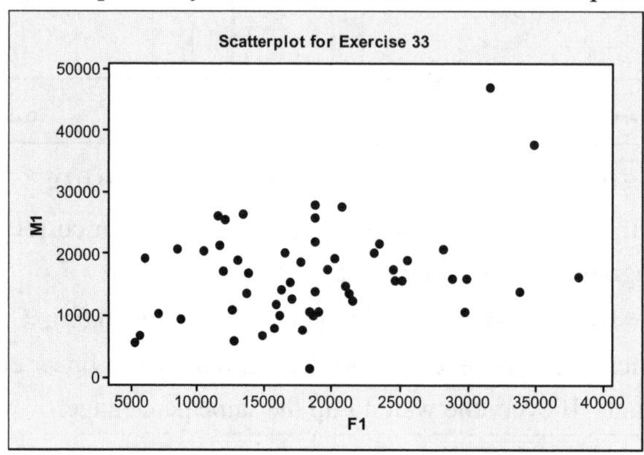

35. a. 0.911
    b. 0.787
    c. 0.9999 (largest)
    d. 0.976
    e. –0.948

y	x	x^2	$\log x$	$\sqrt{x}$	$1/x$
0.3	2	4	0.3010	1.4142	0.5
05	3	9	0.4771	1.7321	0.3333
1.3	20	400	1.3010	4.4721	0.05
1.7	50	2500	1.6990	7.0711	0.02
2.0	95	9025	1.9777	9.7468	0.0105

Section 10-2: Regression

1. a. $\hat{y} = -368 + 130x$
   b. $\hat{y}$ represents the predicted value of price from rating.
3. a. A residual is a value of $y - \hat{y}$, which is the difference between an observed value of y and a predicted value of y.
   b. The regression line has the property that the sum of squares of the residuals is the lowest possible sum.
5. With no significant linear correlation, the best predicted value is $\bar{y} = 5.9$.
7. With a significant linear correlation, the best predicted value is $\hat{y} = -106 + 1.10(180) = 92.0$ kg.

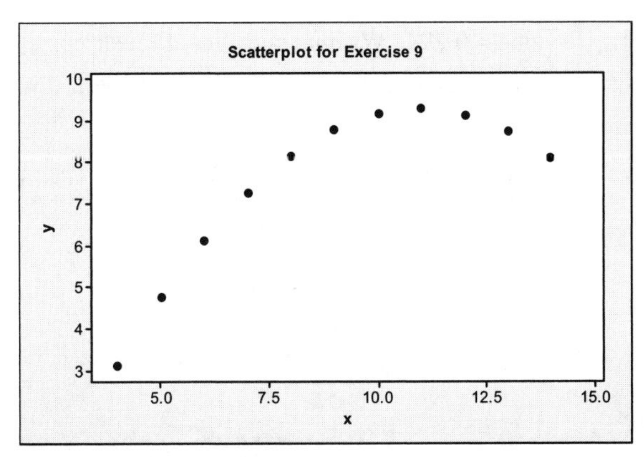

9. $\hat{y} = 3.00 + 0.500x$; The data have a pattern that is not a straight line.

11. a. $\hat{y} = 0.264 + 0.906x$

    b. $\hat{y} = 2 + 0x$ (or $\hat{y} = 2$)

    c. The results are very different, indicating that one point can dramatically affect the regression equation.

13. $\hat{y} = -8.44 + 0.203x$; $r = 0.799$; P-value $= 0.056$; With no significant linear correlation, the best predicted value is $\bar{y} = 5.1$ per 10 million people. The best predicted value is not at all close to the actual Nobel rate of 1.5 per 10 million people.

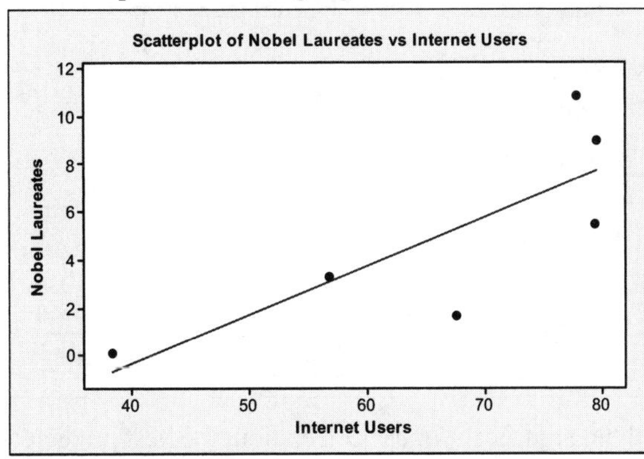

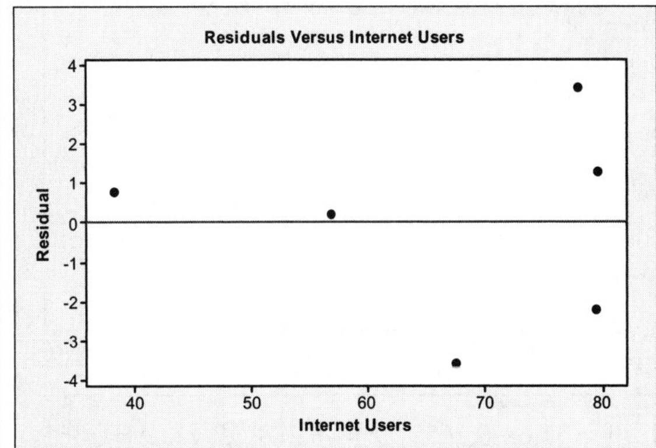

15. $\hat{y} = -0.0111 + 1.01x$; $r = 0.992$; P-value $= 0.000$; With a significant linear correlation, the best predicted value is $\hat{y} = -0.0111 + 1.01(3.00) = \$3.02$. The best predicted subway fare of \$3.02 is not likely to be implemented because it is not a convenient value, such as \$3.00 or \$3.25.

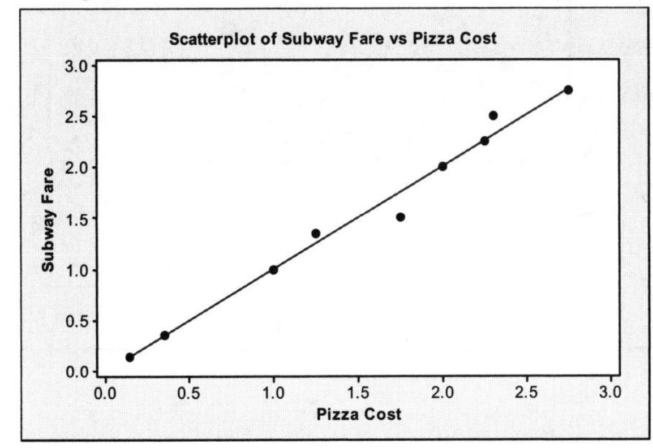

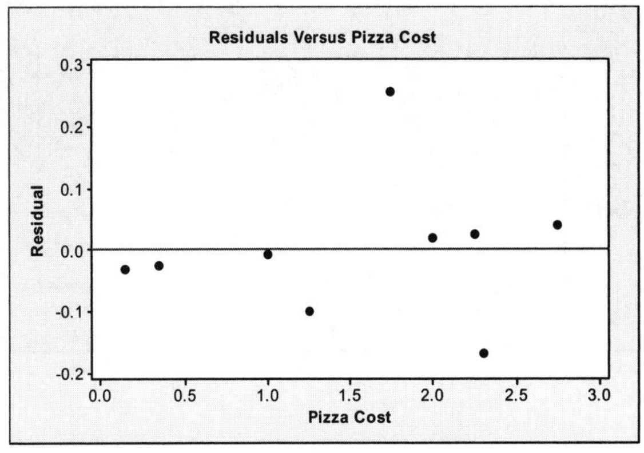

17. $\hat{y} = 125 + 1.73x$; $r = 0.591$; P-value = 0.294; With no significant linear correlation, the best predicted value is $\bar{y} = 177$ cm. Because the best predicted value is the mean height, it would not be helpful to police in trying to obtain a description of the male.

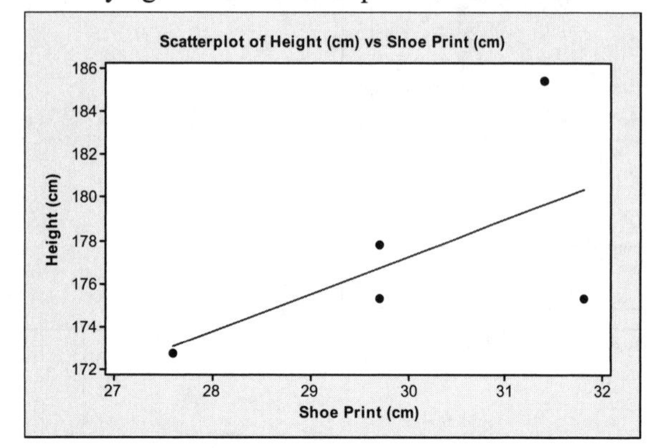

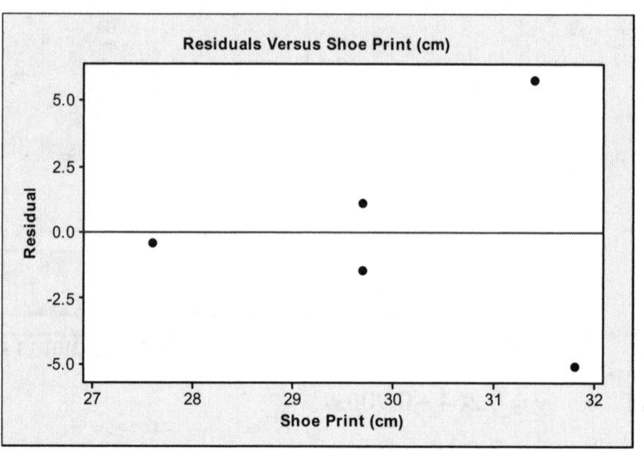

19. $\hat{y} = 16.5 - 0.00282x$; $r = -0.959$; P-value = 0.010; With a significant linear correlation, the best predicted value is $\hat{y} = 16.5 - 0.00282(500) = 15.1$ fatalities per 100,000 population. Common sense suggests that the prediction doesn't make much sense.

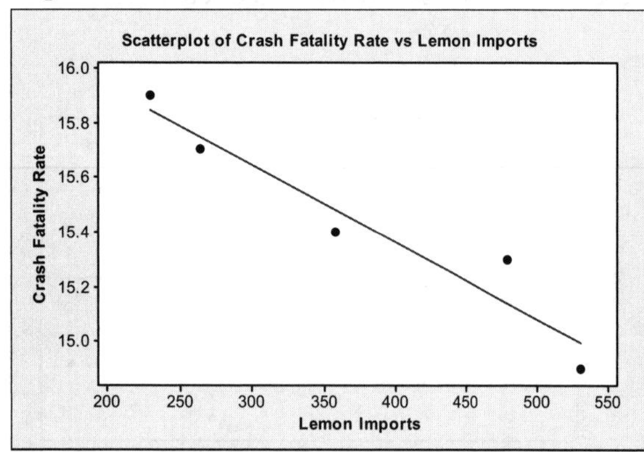

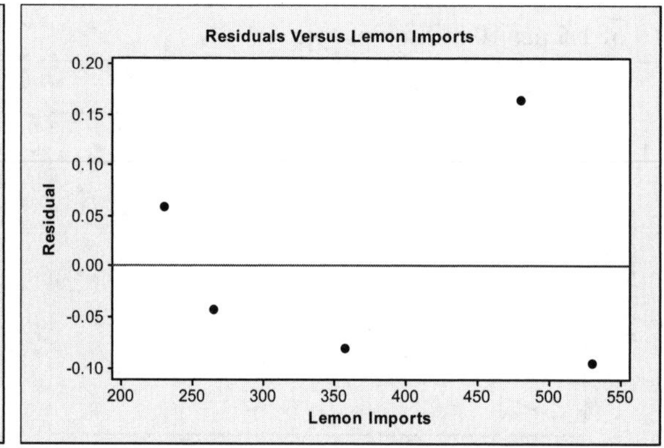

21. $\hat{y} = 51.6 - 0.165x$; $r = -0.288$; P-value = 0.365; With no significant linear correlation, the best predicted value is $\bar{y} = 45$ years, which is not close to the actual age of 33 years.

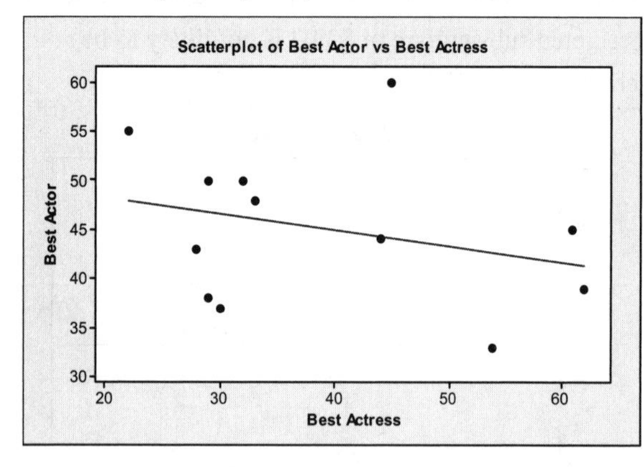

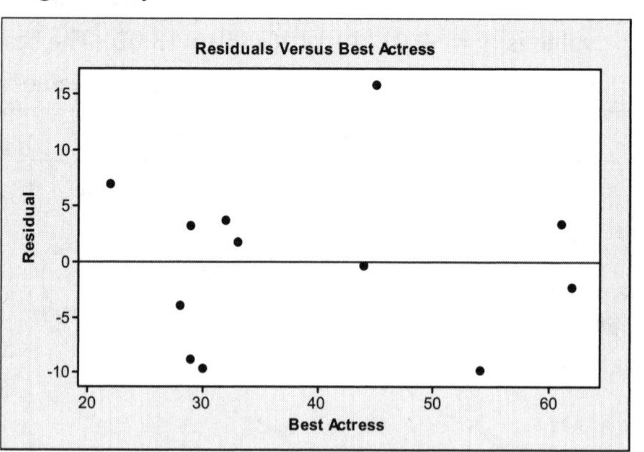

23. $\hat{y} = -157 + 40.2x$; $r = 0.948$; P-value $= 0.004$; With a significant linear correlation, the best predicted value is $\hat{y} = -157 + 40.2(2) = -76.6$ kg; The prediction is a negative weight that cannot be correct. The overhead width of 2 cm is well beyond the scope of the sample widths, so the extrapolation might be off by a considerable amount. Clearly, the predicted negative weight makes no sense.

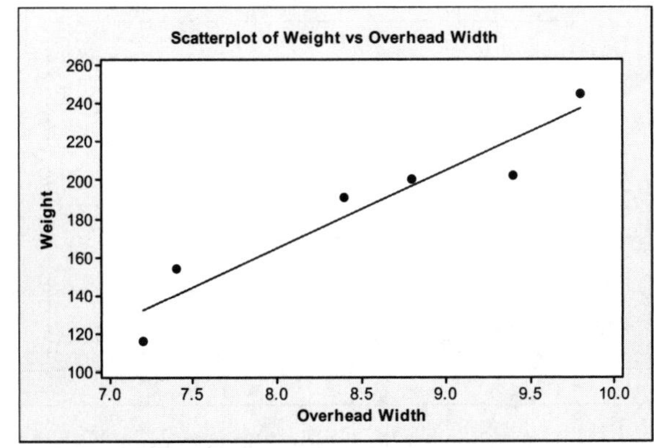

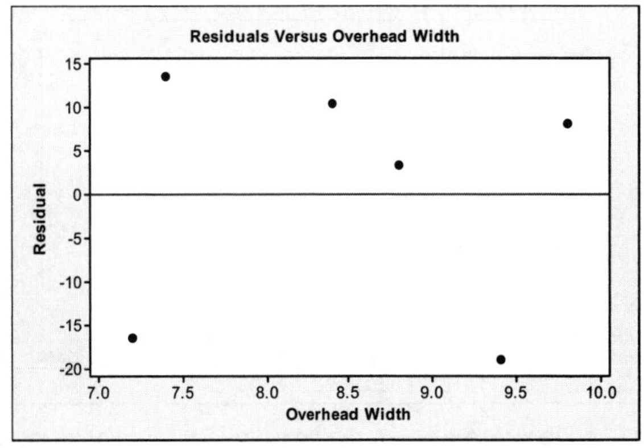

25. $\hat{y} = -0.347 + 0.149x$; $r = 0.828$; P-value $= 0.042$; With a significant linear correlation, the best predicted value is $\hat{y} = -0.347 + 0.149(100) = \$14.55$. Tipping rule: Multiply the bill by 0.149 (or 14.9%) and subtract 35 cents. A more approximate but easier rule is this: Leave a tip of 15%.

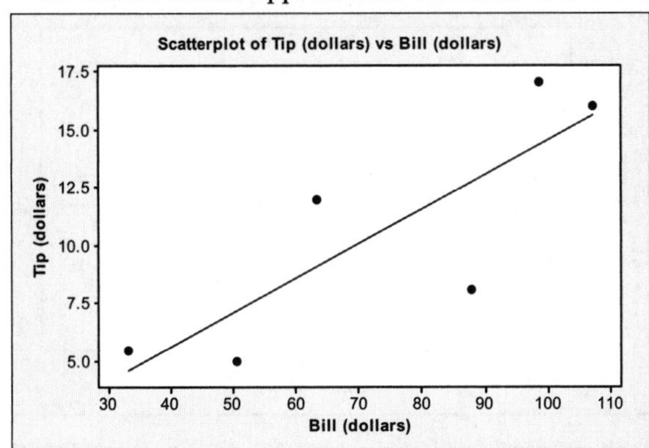

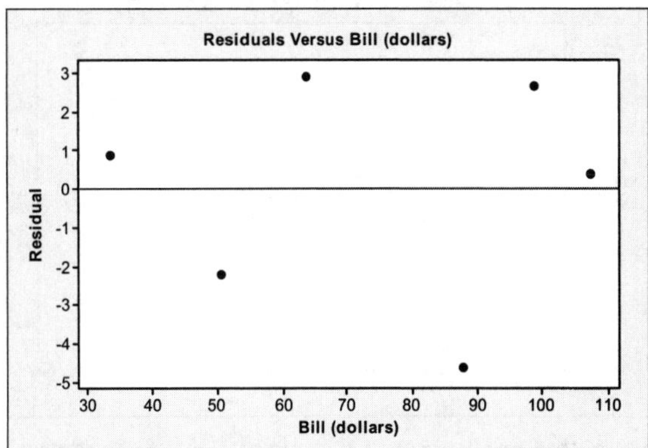

27. $\hat{y} = -0.00396 + 3.14x$; $r = 1.000$; P-value $= 0.000$; With a significant linear correlation, the best predicted value is $\hat{y} = -0.00396 + 3.14(1.50) = 4.7$ cm. Even though the diameter of 1.50 cm is beyond the scope of the sample diameters, the predicted value yields the actual circumference.

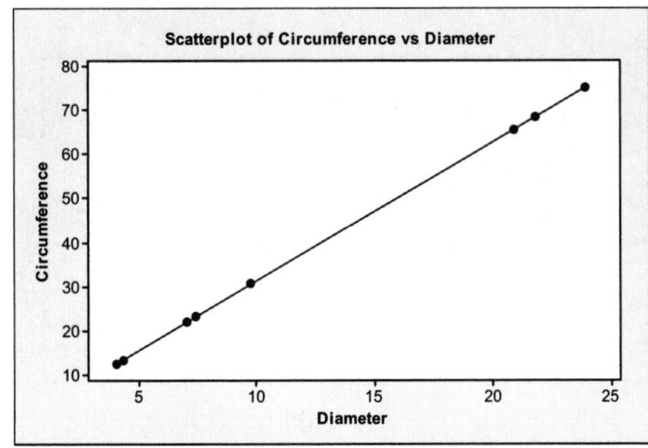

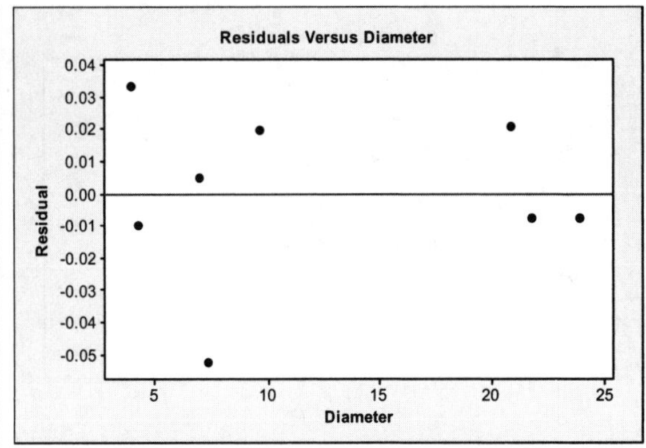

29. $\hat{y} = -23.2 + 0.456x$; $r = 0.702$; P-value $= 0.000$; With a significant linear correlation, the best predicted value is $\hat{y} = -23.2 + 0.456(79.1) = 12.9$ per 10 million people. The best predicted value is not at all close to the actual Nobel rate of 1.5 per 10 million people.

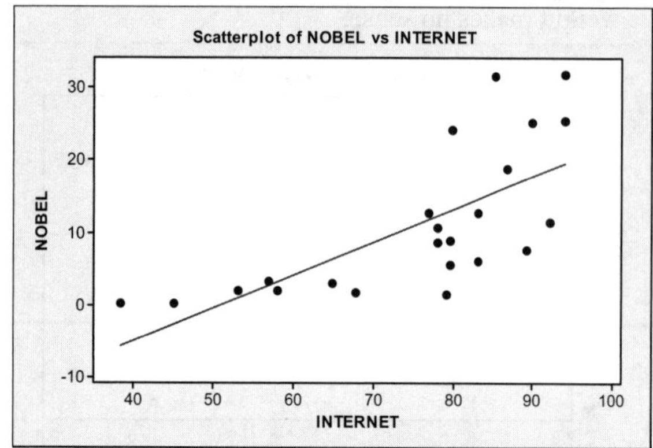

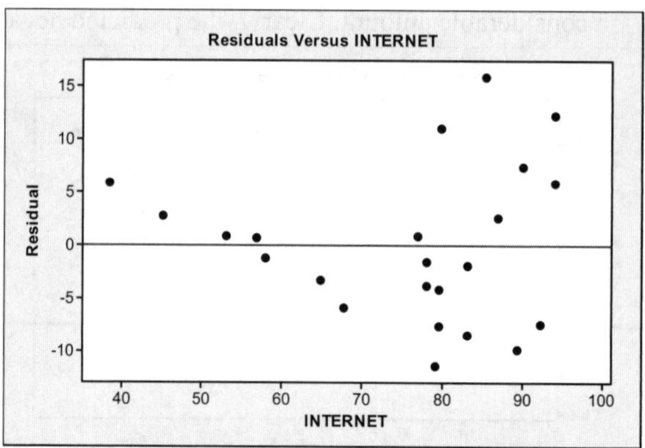

31. $\hat{y} = 93.5 + 2.85x$; $r = 0.594$; P-value $= 0.007$; With a significant linear correlation, the best predicted value is $\hat{y} = 93.5 + 2.85(31.3) = 183$ cm. Although there is a linear correlation, with $r = 0.594$, we see that it is not very strong, so an estimate of the height of a male might be off by a considerable amount.

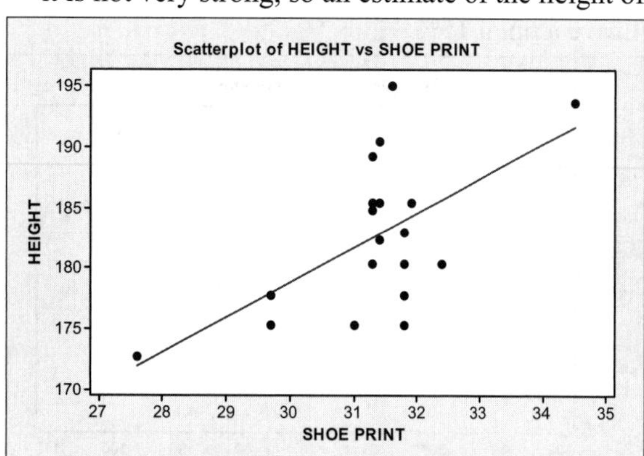

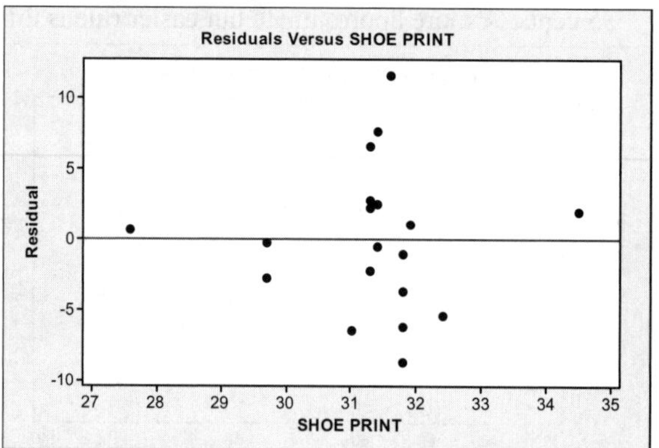

33. $\hat{y} = 13{,}400 + 0.302x$; $r = 0.319$; P-value $= 0.017$; With a significant linear correlation, the best predicted value is $\hat{y} = 13{,}400 + 0.302(16{,}000) = 18{,}200$ words. Although there is a linear correlation, with $r = 0.319$, we see that it is not very close to 1, so an estimate of the number of words spoken by a female might be off by a considerable amount.

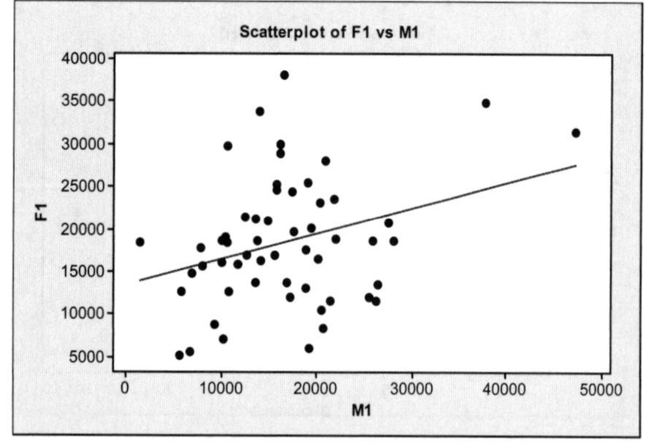

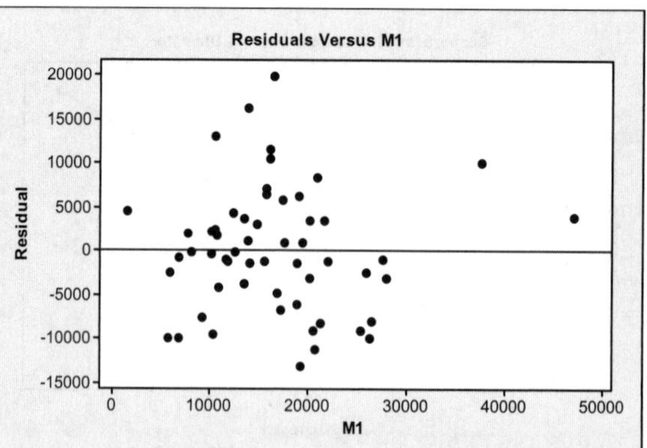

35. a. Using $\hat{y}=-3.37+2.49x$, the sum of squares of the residuals is 823.64.

b. Using $\hat{y}=-3+2.5x$, the sum of squares of the residuals is 827.45, which is larger than 823.64, which is the sum of squares of the residuals for the regression line.

x	$\hat{y}=-3.37+2.49x$	$(\hat{y}-\bar{y})^2$
4.5	7.8350	5.4522
10.2	22.0280	5.1620
4.4	7.5860	1.0282
2.9	3.8510	14.0700
3.9	6.3410	0.0581
0.7	−1.6270	2.9825
8.5	17.7950	56.3250
7.3	14.8070	51.9408
6.3	12.3170	11.0025
11.6	25.5140	164.1986
2.5	2.8550	0.9120
8.8	18.5420	34.1290
3.7	5.8430	6.4668
1.8	1.1120	0.1505
4.5	7.8350	12.7092
9.4	20.0360	29.8553
3.6	5.5940	6.2200
2.0	1.6100	0.0841
3.6	5.5940	15.1632
6.4	12.5660	373.8036
11.9	26.2610	27.4471
9.7	20.7830	3.5457
5.3	9.8270	0.9467
	Sum:	823.65

x	$\hat{y}=-3+2.5x$	$(\hat{y}-\bar{y})^2$
4.5	8.2500	7.5625
10.2	22.5000	3.2400
4.4	8.0000	0.3600
2.9	4.2500	17.2225
3.9	6.7500	0.4225
0.7	−1.2500	1.8225
8.5	18.2500	49.7025
7.3	15.2500	58.5225
6.3	12.7500	14.0625
11.6	26.0000	176.8900
2.5	3.2500	1.8225
8.8	19.0000	39.6900
3.7	6.2500	8.7025
1.8	1.5000	0.0000
4.5	8.2500	9.9225
9.4	20.5000	25.0000
3.6	6.0000	8.4100
2.0	2.0000	0.0100
3.6	6.0000	18.4900
6.4	13.0000	357.2100
11.9	26.7500	22.5625
9.7	21.2500	5.5225
5.3	10.2500	0.3025
	Sum:	827.45

Section 10-3: Prediction Intervals and Variation

1. The value of $s_e=16.27555$ cm is the standard error of estimate, which is a measure of the differences between the observed weights and the weights predicted from the regression equation. It is a measure of the variation of the sample points about the regression line.

3. The coefficient of determination is $r^2=0.155$. We know that 15.5% of the variation in weight is explained by the linear correlation between height and weight, and 84.5% of the variation in weight is explained by other factors and/or random variation.

5. $r^2=(0.874)^2=0.764$; 76.4% of the variation in temperature is explained by the linear correlation between chirps and temperature, and 23.6% of the variation in temperature is explained by other factors and/or random variation.

7. $r^2=(0.885)^2=0.783$; 78.3% of the variation in waist size is explained by the linear correlation between weight and waist size, and 21.7% of the variation in waist size is explained by other factors and/or random variation.

9. $r=0.850$; Critical values $(\alpha=0.05)$: $r=\pm0.404$ (Table: $r\approx\pm0.396$); Yes, there is sufficient evidence to support a claim of a linear correlation between registered boats and manatee fatalities.

11. The best predicted value is 70.5 manatees.

13. 99% CI: 42.7 manatees $< y <$ 98.3 manatees

$\hat{y} = -49.049 + 1.406(85) = 70.46$

$$E = t_{\alpha/2} s_e \sqrt{1 + \frac{1}{n} + \frac{n(x_0 - \bar{x})^2}{n(\Sigma x^2) - (\Sigma x)^2}} = 2.819(9.6605)\sqrt{1 + \frac{1}{24} + \frac{24(85 - 85.25)^2}{24(127822) - (2046)^2}} = 27.79$$

15. 95% CI: 65.1 manatees $< y <$ 106.8 manatees

$\hat{y} = -49.049 + 1.406(96) = 85.93$

$$E = t_{\alpha/2} s_e \sqrt{1 + \frac{1}{n} + \frac{n(x_0 - \bar{x})^2}{n(\Sigma x^2) - (\Sigma x)^2}} = 2.074(9.6605)\sqrt{1 + \frac{1}{24} + \frac{24(96 - 85.25)^2}{24(127822) - (2046)^2}} = 20.42$$

17. a. 10,626.59
    b. 68.83577
    c. 95% CI: 38.0°F $< y <$ 60.4°F

$\hat{y} = 72.5 - 3.68(6.327) = 49.2$

$$E = t_{\alpha/2} s_e \sqrt{1 + \frac{1}{n} + \frac{n(x_0 - \bar{x})^2}{n(\Sigma x^2) - (\Sigma x)^2}} = 2.5706(3.710411)\sqrt{1 + \frac{1}{7} + \frac{7(6.327 - 20.143)^2}{7(3623) - (141)^2}} = 11.2$$

19. a. 352.7278
    b. 109.3722
    c. 90% CI: 71.09°F $< y <$ 88.71°F

$\hat{y} = 27.628 + 0.05227(1000) = 79.90$

$$E = t_{\alpha/2} s_e \sqrt{1 + \frac{1}{n} + \frac{n(x_0 - \bar{x})^2}{n(\Sigma x^2) - (\Sigma x)^2}} = 1.943(4.2695)\sqrt{1 + \frac{1}{8} + \frac{8(1000 - 1016.25)^2}{8(8391204) - (8130)^2}} = 8.81.$$

21. 95% CI: $17.1 < \bar{y} < 26.0$ (values are Nobel Laureates per 10 million people

$\hat{y} = -3.367 + 2.493(10) = 21.565$

$$E = t_{\alpha/2} s_e \sqrt{\frac{1}{n} + \frac{n(x_0 - \bar{x})^2}{n(\Sigma x^2) - (\Sigma x)^2}} = 2.080(6.26267)\sqrt{\frac{1}{23} + \frac{23(10.0 - 5.8)^2}{23(1011.45) - (133.5)^2}} = 4.476$$

Section 10-4: Multiple Regression

1. The response variable is weight and the predictor variables are length and chest size.
3. The unadjusted R^2 increases (or remains the same) as more variables are included, but the adjusted R^2 is adjusted for the number of variables and sample size. The unadjusted R^2 incorrectly suggests that the best multiple regression equation is obtained by including all of the available variables, but by taking into account the sample size and number of predictor variables, the adjusted R^2 is much more helpful in weeding out variables that should not be included.
5. $\text{Son} = 18.0 + 0.504\,\text{Father} + 0.277\,\text{Mother}$
7. A P-value less than 0.0001 is low, but the values of R^2 (0.3649) and adjusted R^2 (0.3552) are not high. Although the multiple regression equation fits the sample data best, it is not a good fit, so it should not be used for predicting the height of a son based on the height of his father and the height of his mother.
9. HWY (highway fuel consumption) because it has the best combination of small P-value (0.000) and highest adjusted R^2 (0.920).

11. CITY $= -3.15 + 0.819$ HWY; That equation has a low P-value of 0.0000 and its adjusted R^2 value of 0.920 isn't very much less than the values of 0.928 and 0.935 that use two predictor variables, so in this case it is better to use the one predictor variable instead of two.
13. The best regression equation is $\hat{y} = 0.127 + 0.0878x_1 - 0.0250x_2$, where x_1 represents tar and x_2 represents carbon monoxide. It is best because it has the highest adjusted R^2 value of 0.927 and the lowest P-value of 0.000. It is a good regression equation for predicting nicotine content because it has a high value of adjusted R^2 and a low P-value. Possible models:

    $\hat{y} = 0.080 + 0.0633x_1$, adjusted $R^2 = 0.877$

    $\hat{y} = 0.328 + 0.0397x_2$, adjusted $R^2 = 0.437$

    $\hat{y} = 0.127 + 0.0878x_1 - 0.0250x_2$, adjusted $R^2 = 0.927$
15. The best regression equation is $\hat{y} = 109 - 0.00670x_1$, where x_1 represents volume. It is best because it has the highest adjusted R^2 value of -0.0513 and the lowest P-value of 0.791. The three regression equations all have adjusted values of R^2 that are very close to 0, so none of them are good for predicting IQ. It does not appear that people with larger brains have higher IQ scores. Possible models:

    $\hat{y} = 109 - 0.00670x_1$, adjusted $R^2 = -0.0513$

    $\hat{y} = 101 - 0.00178x_2$, adjusted $R^2 = -0.0555$

    $\hat{y} = 108 - 0.00694x_1 + 0.00722x_2$, adjusted $R^2 = -0.113$
17. For H_0: $\beta_1 = 0$, Test statistic: $t = \dfrac{0.769317 - 0}{0.0711414} = 10.813917$; P-value < 0.0001; Reject H_0 and conclude that the regression coefficient of $b_1 = 0.769$ should be kept.

    For H_0: $\beta_2 = 0$, Test statistic: $t = \dfrac{1.009510 - 0}{0.0338123} = 29.856$; P-value < 0.0001; Reject H_0 and conclude that the regression coefficient of $b_2 = 1.01$ should be kept.

    It appears that the regression equation should include both independent variables of height and waist circumference.
19. $\hat{y} = 3.06 + 82.4x_1 + 2.91x_2$, where x_1 represents sex and x_2 represents age.

    Female: $\hat{y} = 3.06 + 82.4(0) + 2.91(20) = 61$ lb; Male: $\hat{y} = 3.06 + 82.4(1) + 2.91(20) = 144$ lb; The sex of the bear does appear to have an effect on its weight. The regression equation indicates that the predicted weight of a male bear is about 82 lb more than the predicted weight of a female bear with other characteristics being the same.

Section 10-5: Nonlinear Regression

1. Since the area of a square is the square of its side, the best model quadratic: $y = x^2$; $R^2 = 1$.
3. 25.5% of the variation in Super Bowl points can be explained by the quadratic model that relates the variable of year and the variable of points scored. Because such a small percentage of the variation is explained by the model, the model is not very useful.

5. Quadratic: $d = -4.88t^2 + 0.0214t + 300$

Model	R^2
Linear	0.962
Quadratic	1.000
Logarithmic	0.831
Exponential	0.933
Power	0.783

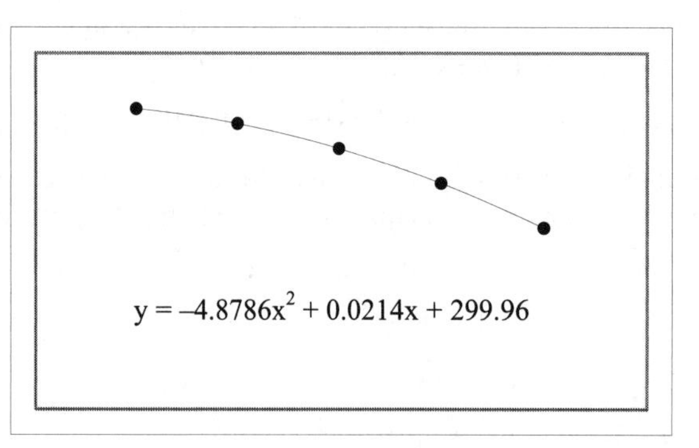

7. Exponential: $y = 1000\left(1.0122^x\right)$; The value of R^2 is slightly higher for the exponential model.

Model	R^2
Linear	0.999
Quadratic	0.999
Logarithmic	0.944
Exponential	0.999
Power	0.973

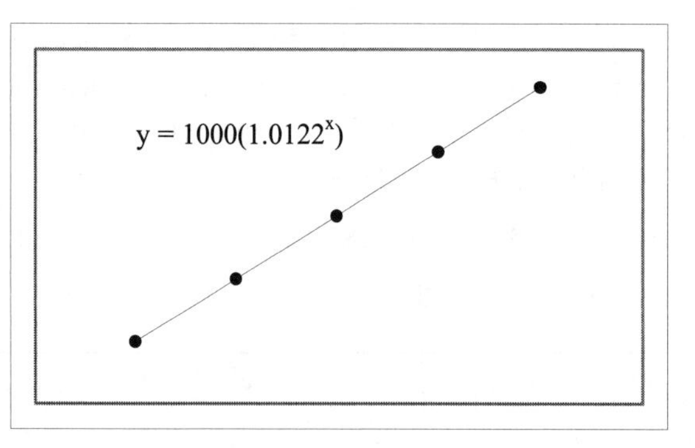

9. Exponential: $y = 10\left(2^x\right)$; with end of first day coded as 1.

Model	R^2
Linear	0.771
Quadratic	0.975
Logarithmic	0.549
Exponential	1.000
Power	0.927

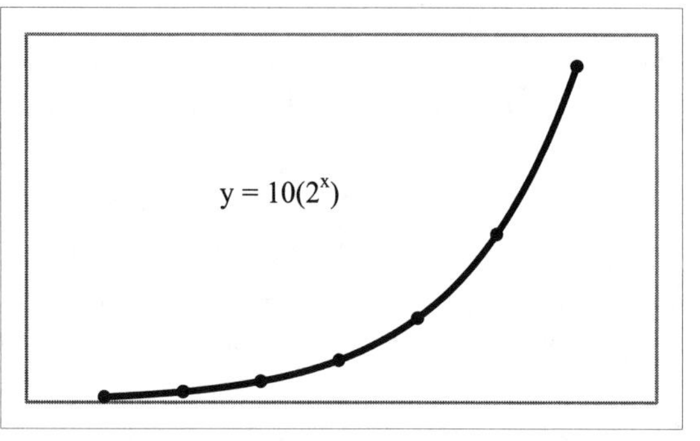

11. Logarithmic: $y = 3.22 + 0.293 \ln x$

Model	R^2
Linear	0.620
Quadratic	0.901
Logarithmic	0.997
Exponential	0.566
Power	0.989

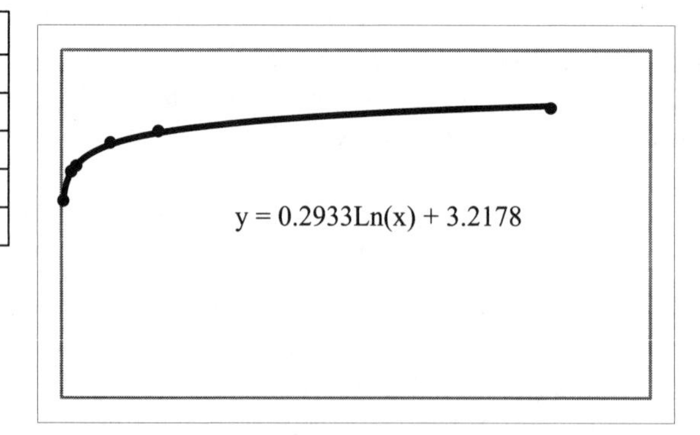

13. Power: $y = 2107.9x^{0.615}$; (Result is based on 1990 coded as 1.) The projected value for 2014 is 15,271 (Using rounded coefficients: 15,261), which is considerably less than the actual value of 18,054.

Modcl	R^2
Linear	0.856
Quadratic	0.876
Logarithmic	0.820
Exponential	0.804
Power	0.896

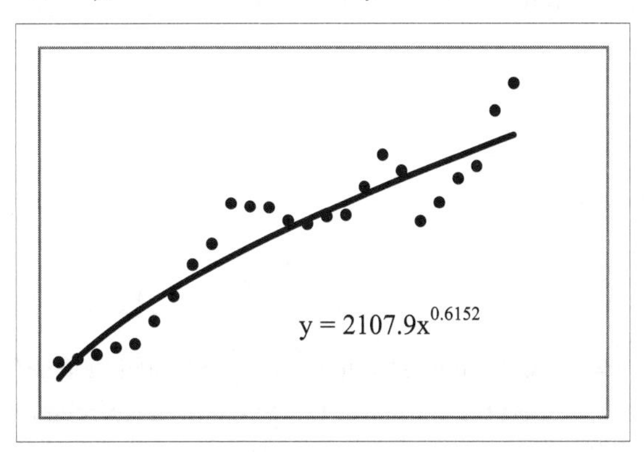

15. Quadratic: $y = 0.700x^2 - 3.41x + 299$; The projected value is $y = 0.700(22)^2 - 3.41(22) + 299 = 563$. (1880–1889 coded as 1.) The decade of 2090–2099 is too far beyond the scope of the available data, so the predicted value is questionable.

Model	R^2
Linear	0.870
Quadratic	0.985
Logarithmic	0.627
Exponential	0.891
Power	0.655

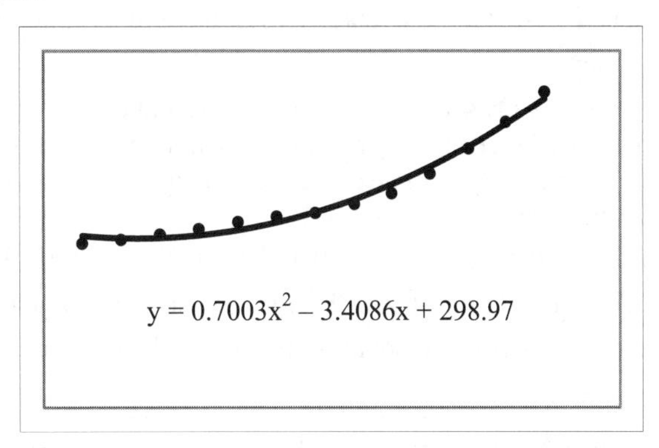

17. a. Exponential: $y = 2^{\frac{2}{3}(x-1)}$ [or $y = 0.629961(1.587401)^x$ for an initial value of 1 that doubles every 1.5 years].

b. Exponential: $y = 1.36558(1.42774)^x$, where x is year after 1970.

Model	R^2
Linear	0.380
Quadratic	0.55
Logarithmic	0.158
Exponential	0.990
Power	0.790

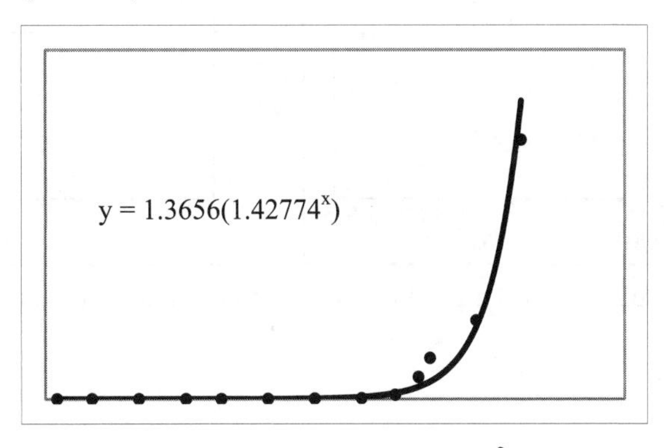

c. Moore's law does appear to be working reasonably well. With $R^2 = 0.990$, the model appears to be very good.

Quick Quiz

1. Conclude that there is not sufficient evidence to support the claim of a linear correlation between enrollment and burglaries.
2. None of the given values change when the variables are switched.
3. No, the value of r does not change if all values of one of the variables are multiplied by the same constant.

4. Because r must be between –1 and 1 inclusive, the value of 1.500 is the result of an error in the calculations.
5. The best predicted number of burglaries is 92.6, which is the mean of the five sample burglary counts.
6. The best predicted number of burglaries would be 123.3, which is found by substituting 50 for x in the regression equation.
7. $r^2 = 0.249$
8. false
9. false
10. $r = -1$

Review Exercises

1. a. $r = 0.962$; P-value $= 0.000$ (Table: P-value < 0.01); Critical values ($\alpha = 0.05$): $r = \pm 0.707$; There is sufficient evidence to support the claim that there is a linear correlation between the amount of tar and the amount of nicotine.
   b. $(0.962)^2 = 0.925$, or 92.5%
   c. $\hat{y} = -0.758 + 0.0920x$
   d. The predicted value is $\hat{y} = -0.758 + 0.0920(23) = 1.358$ mg or 1.4 mg rounded, which is close to the actual amount of 1.3 mg.
2. a. The scatterplot (see part c) shows a pattern with nicotine and carbon monoxide both increasing from left to right, but it is a very weak pattern and the points are not very close to a straight-line pattern, so it appears that there is not sufficient sample evidence to support the claim of a linear correlation between amounts of nicotine and carbon monoxide.
   b. $H_0: \rho = 0$; $H_1: \rho \neq 0$; $r = 0.329$; P-value $= 0.427$ (Table: P-value > 0.05); Critical values ($\alpha = 0.05$): $r = \pm 0.707$; Fail to reject H_0. There is not sufficient evidence to support the claim that there is a linear correlation between amount of nicotine and amount of carbon monoxide.
   c. $\hat{y} = 14.2 + 1.42x$

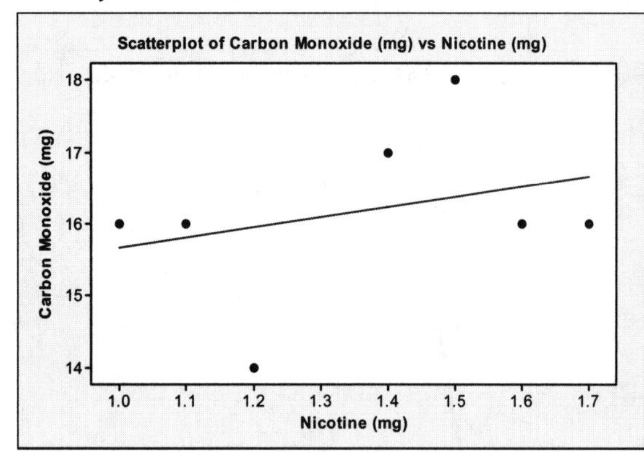

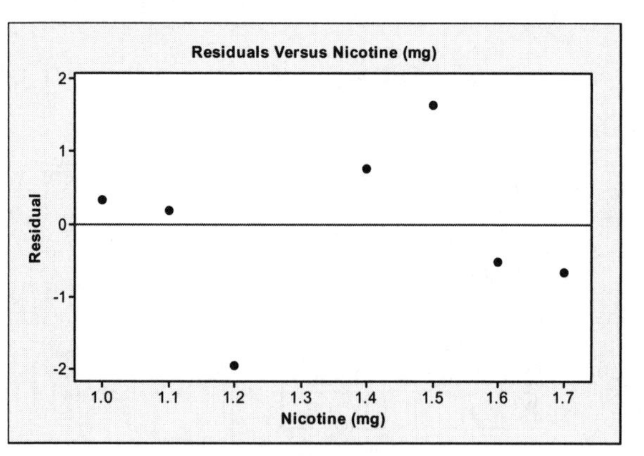

   d. The predicted value is $\bar{y} = 16.1$ mg, which is close to the actual amount of 15 mg.

3. $H_0: \rho = 0$; $H_1: \rho \neq 0$; $r = 0.450$; P-value $= 0.192$ P-value > 0.05); Critical values ($\alpha = 0.05$): $r = \pm 0.632$; Fail to reject H_0. There is not sufficient evidence to support the claim that there is a linear correlation between time and height. Although there is no *linear* correlation between time and height, the scatterplot shows a very distinct pattern revealing that time and height are associated by some function that is not linear.

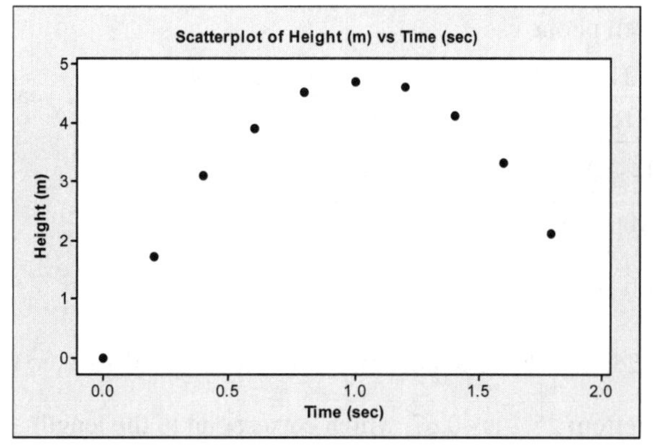

4. a. NICOTINE $= -0.443 + 0.0968\,\text{TAR} - 0.0262\,\text{CO}$, or $\hat{y} = -0.443 + 0.0968x_1 - 0.0262x_2$, where x_1 represents tar and x_2 represents carbon monoxide.
   b. $R^2 = 0.936$; adjusted $R^2 = 0.910$; P-value $= 0.001$
   c. With high values of R^2 and adjusted R^2 and a small P-value of 0.001, it appears that the regression equation can be used to predict the amount of nicotine given the amounts of tar and carbon monoxide.
   d. The predicted value is $\hat{y} = -0.443 + 0.0968(23) - 0.0262(15) = 1.39$ mg or 1.4 mg rounded, which is close to the actual value of 1.3 mg of nicotine.

Cumulative Review Exercises

1. a. $\bar{x} = 35.91$, $Q_2 = 36.10$, range $= 76.40$, $s = 31.45$, $s^2 = 989.10$
   b. quantitative data
   c. ratio
2. $r = 0.731$; P-value $= 0.039$ (Table: P-value < 0.05); Critical values: $r = \pm 0.707$; There is sufficient evidence to support the claim of a linear correlation between the DJIA values and sunspot numbers. Because it would be reasonable to think that there is no correlation between stocks and sunspot numbers, the result is not as expected. Although there appears to be a linear correlation, a reasonable investor would be wise to ignore sunspot numbers when investing in stocks.
3. The highest sunspot number is 79.3, which converts to $z = (79.3 - 35.91)/31.45 = 1.38$. The highest sunspot number is not significantly high because its z score of 1.38 shows that it is within 2 standard deviations of the mean.
4. The data do not appear to fit the loose definition of a normal distribution and $n < 30$, so proceed with caution. $H_0: \mu = 49.7$; $H_1: \mu \neq 49.7$;

   Test statistic: $t = \dfrac{35.91 - 49.7}{31.45/\sqrt{8}} = -1.240$; Critical values: $t = \pm 2.365$; P-value $= 0.255$

   (Table: P-value > 0.20); Fail to reject H_0. There is not sufficient evidence to warrant rejection of the claim that the sample is from a population with a mean equal to 49.7.
5. 95% CI: $\bar{x} \pm t_{\alpha/2}\dfrac{s}{\sqrt{n}} = 35.91 \pm 2.365 \cdot \dfrac{31.45}{\sqrt{8}} \Rightarrow 9.62 < \mu < 62.21$; We have 95% confidence that the interval limits of 9.62 and 62.21 contain the true value of the mean of the population of sunspot numbers.

6. $H_0: p = 0.10$; $H_1: p < 0.10$; Test statistic: $z = \dfrac{0.07 - 0.10}{\sqrt{\dfrac{(0.10)(0.90)}{2733}}} = -5.23$ ($z = -5.25$ using $x = 191$);

   P-value $= P(z < -5.23) = 0.0000$ (Table: 0.0001); Critical value: $z = -1.645$;

   Reject H_0. There is sufficient evidence to support the claim that fewer than 10% of police traffic stops are attributable to improper cell phone use.

7. $\bar{x} = \dfrac{7 \cdot 6.5 + 15 \cdot 14.5 + 19 \cdot 21.0 + 19 \cdot 31.0 + 15 \cdot 44.5 + 11 \cdot 54.5 + 14 \cdot 70}{7 + 15 + 19 + 19 + 15 + 11 + 14} = 35.2$ years

   $s = \sqrt{\dfrac{100\left(7 \cdot 6.5^2 + \cdots + 14 \cdot 70^2\right) - \left(7 \cdot 6.5 + \cdots + 14 \cdot 70\right)^2}{100(100-1)}} = 19.7$ years

   $s^2 = (19.7 \text{ years})^2 = 389.6 \text{ years}^2$

8. a. $z_{x=30} = \dfrac{30-35}{20} = -0.25$; which has a probability of 0.4013, or 40.13%, to the left.

   b. The z score for the bottom 25% is –0.67, which correspond to the length $-0.67 \cdot 20 + 35 = 21.6$ years (Tech: 21.5 years).

   c. $z_{x=30} = \dfrac{30-35}{20/\sqrt{25}} = -1.25$; which has a probability of 0.1056 to the left.

   d. 0+ or 0.0000 (from $0.4013^{25} = 0.0000$); The audience for a particular movie and showtime is not a simple random sample. Some movies and showtimes attract very young audiences.

Chapter 11: Goodness-of-Fit and Contingency Tables

Section 11-1: Goodness-of-Fit

1. a. Observed values are represented by O and expected values are represented by E.
   b. For the leading digit of 2, $O = 62$ and $E = (317)(0.176) = 55.792$.
   c. For the leading digit of 2, $(O - E)^2/E = 0.691$.

3. There is sufficient evidence to warrant rejection of the claim that the leading digits have a distribution that fits well with Benford's law.

5. H_0: The frequency counts agree with the claimed distribution.
   H_1: The frequency counts do not agree with the claimed distribution.

   Test statistic: $\chi^2 = 8.185$; P-value $= 0.516$ (Table: P-value > 0.10); Critical value: $\chi^2 = 16.919$; Fail to reject H_0. There is not sufficient evidence to warrant rejection of the claim that the observed outcomes agree with the expected frequencies. The slot machine appears to be functioning as expected.

7. $H_0: p_1 = p_2 = p_3 = p_4 = p_5 = p_6 = 1/6$;
   H_1: At least one of the proportions is not equal to the given claimed value.

   Test statistic: $\chi^2 = 5.860$; P-value $= 0.320$ (Table: P-value > 0.10); Critical value: $\chi^2 = 11.071$; Fail to reject H_0. There is not sufficient evidence to support the claim that the outcomes are not equally likely. The outcomes appear to be equally likely, so the loaded die does not appear to behave differently from a fair die.

$$\chi^2 = \frac{(27-28.571)^2}{200/7} + \frac{(31-28.571)^2}{200/7} + \cdots + \frac{(28-28.571)^2}{200/7} + \frac{(32-28.571)^2}{200/7} = 5.860 \text{ (df} = 5)$$

9. $H_0: p_{YS} = 9/16, p_{GS} = 3/16, p_{YW} = 3/16, p_{GW} = 1/16$;
   H_1: At least one of the proportions is not equal to the given claimed value.

   Test statistic: $\chi^2 = 11.161$; P-value $= 0.011$ (Table: P-value < 0.025); Critical value: $\chi^2 = 7.815$; Reject H_0. There is sufficient evidence to support the claim that the results contradict Mendel's theory.

$$\chi^2 = \frac{(307-281.25)^2}{0.5625 \cdot 500} + \frac{(77-93.75)^2}{0.1875 \cdot 500} + \frac{(98-93.75)^2}{0.1875 \cdot 500} + \frac{(18-32.25)^2}{0.0625 \cdot 500} = 11.161 \text{ (df} = 3)$$

11. $H_0: p_{Sun} = p_{Mon} = p_{Tue} = p_{Wed} = p_{Thu} = p_{Fri} = p_{Sat} = 1/7$;
    H_1: At least one of the proportions is not equal to the given claimed value.

    Test statistic: $\chi^2 = 29.814$; P-value $= 0.0000$ (Table: P-value < 0.005); Critical value: $\chi^2 = 16.812$; Reject H_0. There is sufficient evidence to warrant rejection of the claim that the different days of the week have the same frequencies of police calls. The highest numbers of calls appear to fall on Friday and Saturday, and these are weekend days with disproportionately more partying and drinking.

$$\chi^2 = \frac{(130-156.43)^2}{1095/7} + \frac{(114-156.43)^2}{1095/7} + \cdots + \frac{(179-156.43)^2}{1095/7} + \frac{(196-156.43)^2}{1095/7} = 29.814 \text{ (df} = 6)$$

13. $H_0: p_1 = p_2 = p_3 = p_4 = p_5 = p_6 = p_7 = p_8 = p_9 = p_{10} = 0.1$;
    H_1: At least one of the proportions is not equal to the given claimed value.

    Test statistic: $\chi^2 = 13.855$; P-value $= 0.128$ (Table: P-value > 0.10); Critical value: $\chi^2 = 16.919$; Fail to reject H_0. There is not sufficient evidence to warrant rejection of the claim that the likelihood of winning is the same for the different post positions. Based on these results, post position should not be considered when betting on the Kentucky Derby race.

$$\chi^2 = \frac{(19-11.7)^2}{0.1 \cdot 117} + \frac{(14-11.7)^2}{0.1 \cdot 117} + \cdots + \frac{(5-11.7)^2}{0.1 \cdot 117} + \frac{(11-11.7)^2}{0.1 \cdot 117} = 13.855 \text{ (df} = 9)$$

15. $H_0: p_4 = 0.125, p_5 = 0.25, p_6 = 0.3125, p_7 = 0.3125$;
    H_1: At least one of the proportions is not equal to the given claimed value.

    Test statistic: $\chi^2 = 8.882$; P-value $= 0.031$ (Table: P-value < 0.05); Critical value: $\chi^2 = 7.815$; Reject H_0. There is sufficient evidence to warrant rejection of the claim that the actual numbers of games fit the distribution indicated by the proportions listed in the given table.

$$\chi^2 = \frac{(21-13.125)^2}{0.125 \cdot 105} + \frac{(23-26.25)^2}{0.25 \cdot 105} + \frac{(23-32.813)^2}{0.3125 \cdot 105} + \frac{(38-32.813)^2}{0.3125 \cdot 105} = 8.882 \text{ (df} = 3)$$

17. $H_0: p_{\text{Sun}} = p_{\text{Mon}} = p_{\text{Tue}} = p_{\text{Wed}} = p_{\text{Thu}} = p_{\text{Fri}} = p_{\text{Sat}} = 1/7$;
    H_1: At least one of the proportions is not equal to the given claimed value.

    Test statistic: $\chi^2 = 9.500$; P-value $= 0.147$ (Table: P-value > 0.10); Critical value: $\chi^2 = 16.812$; Fail to reject H_0. There is not sufficient evidence to support the claim that births do not occur on the seven different days of the week with equal frequency.

Day	**Sun**	**Mon**	**Tue**	**Wed**	**Thu**	**Fri**	**Sat**
Expected	57.14	57.14	57.14	57.14	57.14	57.14	57.14
Observed	53	52	66	72	57	57	43

$$\chi^2 = \frac{(53-57.14)^2}{400/7} + \frac{(52-57.14)^2}{400/7} + \cdots + \frac{(57-57.14)^2}{400/7} + \frac{(43-57.14)^2}{400/7} = 9.500 \text{ (df} = 6)$$

19. $H_0: p_{\text{red}} = 0.13, p_{\text{orange}} = 0.20, p_{\text{yellow}} = 0.14, p_{\text{brown}} = 0.13, p_{\text{blue}} = 0.24, p_{\text{green}} = 0.16$;
    H_1: At least one of the proportions is not equal to the given claimed value.

    Test statistic: $\chi^2 = 6.682$; P-value $= 0.245$ (Table: P-value > 0.10); Critical value: $\chi^2 = 11.071$; Fail to reject H_0. There is not sufficient evidence to warrant rejection of the claim that the color distribution is as claimed.

Color	**Red**	**Orange**	**Yellow**	**Brown**	**Blue**	**Green**
Expected	13	20	14	13	24	16
Observed	13	25	8	8	27	19

$$\chi^2 = \frac{(13-13)^2}{13} + \frac{(20-25)^2}{20} + \frac{(14-8)^2}{8} + \frac{(24-27)^2}{24} + \frac{(16-19)^2}{16} = 6.682 \text{ (df} = 5)$$

21. $H_0: p_1 = 0.301, p_2 = 0.176, p_3 = 0.125, \ldots, p_7 = 0.058, p_8 = 0.051, p_9 = 0.046$;
    H_1: At least one of the proportions is not equal to the given claimed value.

    Test statistic: $\chi^2 = 3650.251$; P-value $= 0.000$ (Table: P-value < 0.005); Critical value: $\chi^2 = 20.090$; Reject H_0. There is sufficient evidence to warrant rejection of the claim that the leading digits are from a population with a distribution that conforms to Benford's law. It does appear that the checks are the result of fraud (although the results cannot confirm that fraud is the cause of the discrepancy between the observed results and the expected results).

$$\chi^2 = \frac{(0-235.98)^2}{0.301 \cdot 784} + \frac{(15-137.98)^2}{0.176 \cdot 784} + \cdots + \frac{(23-39.98)^2}{0.051 \cdot 784} + \frac{(0-36.06)^2}{0.046 \cdot 784} = 3650.251 \text{ (df} = 8)$$

23. H_0: $p_1 = 0.301, p_2 = 0.176, p_3 = 0.125, \ldots, p_7 = 0.058, p_8 = 0.051, p_9 = 0.046$;
H_1: At least one of the proportions is not equal to the given claimed value.

Test statistic: $\chi^2 = 1.762$; P-value $= 0.988$ (Table: P-value > 0.10); Critical value: $\chi^2 = 15.507$; Fail to reject H_0. There is not sufficient evidence to warrant rejection of the claim that the leading digits are from a population with a distribution that conforms to Benford's law. The tax entries do appear to be legitimate.

$$\chi^2 = \frac{(152-153.8)^2}{0.301 \cdot 511} + \frac{(89-89.9)^2}{0.176 \cdot 511} + \cdots + \frac{(25-26.1)^2}{0.051 \cdot 511} + \frac{(27-23.5)^2}{0.046 \cdot 511} = 1.762 \text{ (df = 8)}$$

25. H_0: Heights selected from a normal distribution.
H_1: Heights not selected from a normal distribution.

	Height (cm)	Less than 155.45	155.45 – 162.05	162.05–168.65	Greater than 168.65
a.	**Frequency**	26	46	49	26
b.	**Tech:**	0.2023	0.3171	0.3046	0.1761
	Table:	0.2033	0.3166	0.3039	0.1762
c.	**Tech:**	0.2023(247) = 29.7381	0.3171(247) = 46.6137	0.3046(247) = 44.7762	0.1761(247) = 25.8867
	Table:	0.2033(247) = 29.8851	0.3166(247) = 46.5402	0.3039(247) = 44.6733	0.1762(247) = 25.9014

d. Test statistic: $\chi^2 = 0.877$ (Table: $\chi^2 = 0.831$); P-value $= 0.831$ (Table: P-value > 0.10); Critical value: $\chi^2 = 11.345$; Fail to reject H_0. There is not sufficient evidence to warrant rejection of the claim that heights were randomly selected from a normally distributed population. The test suggests that we cannot rule out the possibility that the data are from a normally distributed population.

$$\chi^2 = \frac{(26-29.7381)^2}{29.7381} + \frac{(46-46.6137)^2}{46.6137} + \frac{(49-44.7762)^2}{44.7762} + \frac{(26-25.8867)^2}{25.8867} = 0.877 \text{ (df = 3)}$$

Section 11-2: Contingency Tables

1. a. $E = \frac{(16+50+3)(40+3)}{436+166+40+16+50+3} = \frac{(69)(43)}{711} = 4.173$

b. Because the expected frequency of a cell is less than 5, the requirements for the hypothesis test are not satisfied.

3. Test statistic: $\chi^2 = 64.517$; P-value $= 0.0000$; Reject the null hypothesis of independence between handedness and cell phone ear preference.

5. H_0: Whether a subject lies is independent of polygraph indication.
H_1: Subject lies depends on polygraph indication.

Test statistic: $\chi^2 = 25.571$; P-value $= 0.0000$ (Table: P-value < 0.005); Critical value: $\chi^2 = 3.841$; df $= (2-1)(2-1) = 1$; Reject H_0. There is sufficient evidence to warrant rejection of the claim that whether a subject lies is independent of the polygraph test indication. The results suggest that polygraphs are effective in distinguishing between truths and lies, but there are many false positives and false negatives, so they are not highly reliable.

$$\chi^2 = \frac{(15-27.3)^2}{27.3} + \frac{(42-29.7)^2}{29.7} + \frac{(32-19.7)^2}{19.7} + \frac{(9-21.3)^2}{21.3} = 25.571$$

7. H_0: Whether texting while driving is independent of driving when drinking alcohol.
   H_1: Texting while driving depends on driving when drinking alcohol.

   Test statistic: $\chi^2 = 576.224$; P-value $= 0.0000$ (Table: P-value < 0.005); Critical value: $\chi^2 = 3.841$; df $= (2-1)(2-1) = 1$; Reject H_0. There is sufficient evidence to warrant rejection of the claim that success is independent of the type of treatment. The results suggest that the surgery treatment is better.

$$\chi^2 = \frac{(731-394.7)^2}{394.7} + \frac{(3054-3390.3)^2}{3390.3} + \frac{(156-492.3)^2}{492.3} + \frac{(4564-4227.7)^2}{4227.7} = 576.224$$

9. H_0: Whether students spent or kept the money is independent of form of \$1.
   H_1: Students spending or keeping the money depends on form of \$1.

   Test statistic: $\chi^2 = 12.162$; P-value $= 0.001$ (Table: P-value < 0.005); Critical value: $\chi^2 = 3.841$; df $= (2-1)(2-1) = 1$; Reject H_0. There is sufficient evidence to warrant rejection of the claim that whether students purchased gum or kept the money is independent of whether they were given four quarters or a \$1 bill. It appears that there is a denomination effect.

$$\chi^2 = \frac{(27-18.8)^2}{18.8} + \frac{(16-24.2)^2}{24.2} + \frac{(12-20.2)^2}{20.2} + \frac{(34-25.8)^2}{25.8} = 12.162$$

11. H_0: Whether gender is independent of whether call is overturned.
    H_1: Call is overturned depends on gender.

    Test statistic: $\chi^2 = 0.064$; P-value $= 0.801$ (Table: P-value > 0.10); Critical value: $\chi^2 = 3.841$; df $= (2-1)(2-1) = 1$; Fail to reject H_0. There is not sufficient evidence to warrant rejection of the claim that the gender of the tennis player is independent of whether the call is overturned. Neither men nor women appear to be better at challenging calls.

$$\chi^2 = \frac{(161-162.5)^2}{162.5} + \frac{(376-374.6)^2}{374.6} + \frac{(69-66.6)^2}{66.6} + \frac{(152-135.5)^2}{135.5} = 0.064\,11.$$

13. H_0: Whether direction of kick is independent of direction of the goalkeeper jump.
    H_1: Direction of kick depends on direction of the goalkeeper jump.

    Test statistic: $\chi^2 = 14.589$; P-value $= 0.0056$ (Table: P-value < 0.01); Critical value: $\chi^2 = 9.488$; df $= (3-1)(3-1) = 4$; Reject H_0. There is sufficient evidence to warrant rejection of the claim that the direction of the kick is independent of the direction of the goalkeeper jump. The results do not support the theory that because the kicks are so fast, goalkeepers have no time to react. It appears that goalkeepers can choose directions based on the directions of the kicks.

$$\chi^2 = \frac{(54-45.4)^2}{45.4} + \frac{(1-5.8)^2}{5.8} + \frac{(37-40.9)^2}{40.9} + \frac{(41-40.4)^2}{40.4} + \frac{(10-5.2)^2}{5.2}$$
$$+ \frac{(31-36.4)^2}{36.4} + \frac{(46-55.2)^2}{55.2} + \frac{(7-7.1)^2}{7.1} + \frac{(59-49.7)^2}{49.7} = 14.589$$

15. H_0: Whether getting a cold is independent of treatment.
    H_1: Getting a cold depends on treatment.

    Test statistic: $\chi^2 = 2.925$; P-value $= 0.232$ (Table: P-value > 0.10); Critical value: $\chi^2 = 5.991$; df $= (2-1)(3-1) = 2$; Fail to reject H_0. There is not sufficient evidence to warrant rejection of the claim that getting a cold is independent of the treatment group. The results suggest that echinacea is not effective for preventing colds.

$$\chi^2 = \frac{(88-88.6)^2}{88.6} + \frac{(48-44.7)^2}{44.7} + \frac{(42-44.7)^2}{44.7} + \frac{(15-14.4)^2}{14.4} + \frac{(4-7.3)^2}{7.3} + \frac{(10-7.3)^2}{7.3} = 2.925$$

17. H_0: Whether cooperation of the subject is independent of age category.

    H_1: Cooperation of the subject depends on age category.

    Test statistic: $\chi^2 = 20.271$; P-value $= 0.0011$ (Table: P-value < 0.005); Critical value: $\chi^2 = 15.086$; df $= (2-1)(6-1) = 5$; Reject H_0. There is sufficient evidence to warrant rejection of the claim that cooperation of the subject is independent of the age category. The age group of 60 and over appears to be particularly uncooperative.

$$\chi^2 = \frac{(73-73.1)^2}{73.1} + \cdots + \frac{(202-218.5)^2}{218.5} + \frac{(11-10.9)^2}{10.9} + \cdots + \frac{(49-32.5)^2}{32.5} = 20.271$$

19. H_0: Whether state is independent of car having front and rear license plates.

    H_1: State depends on car having front and rear license plates.

    Test statistic: $\chi^2 = 50.446$; P-value $= 0.0000$ (Table: P-value < 0.005); Critical value: $\chi^2 = 5.991$; df $= (2-1)(3-1) = 2$; Reject H_0. There is sufficient evidence to warrant rejection of the claim of independence between the state and whether a car has front and rear license plates. It does not appear that the license plate laws are followed at the same rates in the three states.

$$\chi^2 = \frac{(35-34.6)^2}{34.6} + \cdots + \frac{(9-33.8)^2}{33.8} + \frac{(528-528.4)^2}{528.4} + \cdots + \frac{(541-516.2)^2}{516.2} = 50.446$$

21. From Exercise 9, $\chi^2 = 12.1619258$ and from Review Exercise 1 in Chapter 9, $z = 3.487395274$, so $z^2 = \chi^2$. The critical values are : $\chi^2 = 3.841$ and $z^2 = \pm 1.96$, so $z^2 = \chi^2$ (approximately).

Quick Quiz

1. $H_0: p_0 = p_1 = p_2 = p_3 = p_4 = p_5 = p_6 = p_7 = p_8 = p_9 = 0.1$;

   H_1: At least one of the probabilities is different from the others.

2. $O = 27$ and $E = 0.1 \cdot 300 = 30$
3. right-tailed
4. df $= n - 1 = 10 - 1 = 9$
5. There is not sufficient evidence to warrant rejection of the claim that the last digits are equally likely. Because reported heights would likely include more last digits of 0 and 5, it appears that the heights were measured instead of reported. (Also, most U.S. residents would have difficulty reporting heights in centimeters, because the United States, Liberia, and Myanmar are the only countries that continue to use the Imperial system of measurement.)
6. H_0: Surviving the sinking is independent of whether the person is a man, woman, boy, or girl.

   H_1: Surviving the sinking and whether the person is a man, woman, boy, or girl are somehow related.

7. chi-squared distribution
8. right-tailed
9. df $= (r-1)(c-1) = (2-1)(4-1) = 3$
10. There is sufficient evidence to warrant rejection of the claim that surviving the sinking is independent of whether the person is a man, woman, boy, or girl. Most of the women survived, 45% of the boys survived, and most girls survived, but only about 20% of the men survived, so it appears that the rule was followed quite well.

Review Exercises

1. H_0: $p_{\text{Sun}} = p_{\text{Mon}} = p_{\text{Tue}} = p_{\text{Wed}} = p_{\text{Thu}} = p_{\text{Fri}} = p_{\text{Sat}} = 1/7$;
   H_1: At least one of the proportions is not equal to the given claimed value.
   Test statistic: $\chi^2 = 787.018$; P-value $= 0.0000$ (Table: P-value < 0.005); Critical value: $\chi^2 = 16.812$; Reject H_0. There is sufficient evidence to warrant rejection of the claim that auto fatalities occur on the different days of the week with the same frequency. Because people generally have more free time on weekends and more drinking occurs on weekends, the days of Friday, Saturday, and Sunday appear to have disproportionately more fatalities.
   $$\chi^2 = \frac{(5304-4674.14)^2}{32{,}719/7} + \cdots + \frac{(5985-4674.14)^2}{32{,}719/7} = 787.018 \text{ (df} = 6)$$
2. H_0: Whether health condition is independent of type of filling.
   H_1: Health condition depends on type of filling.
   Test statistic: $\chi^2 = 0.751$; P-value $= 0.356$ (Table: P-value > 0.10); Critical value: $\chi^2 = 3.841$; $\text{df} = (2-1)(2-1) = 1$; Fail to reject H_0. There is not sufficient evidence to warrant rejection of the claim of independence between the type of filling and adverse health conditions. Fillings that contain mercury do not appear to affect health conditions.
   $$\chi^2 = \frac{(135-140)^2}{140} + \frac{(145-140)^2}{140} + \frac{(132-127)^2}{127} + \frac{(122-127)^2}{127} = 0.751$$
3. H_0: The frequency counts agree with the claimed distribution.
   H_1: The frequency counts do not agree with the claimed distribution.
   Test statistic: $\chi^2 = 5.624$; P-value $= 0.467$ (Table: P-value > 0.10); Critical value: $\chi^2 = 12.592$; Fail to reject H_0. There is not sufficient evidence to warrant rejection of the claim that the actual eliminations agree with the expected numbers. The leadoff singers do appear to be at a disadvantage because 20 of them were eliminated compared to the expected value of 12.9 eliminations, but that result does not appear to be *significantly* high.
   $$\chi^2 = \frac{(20-12.9)^2}{12.9} + \frac{(12-12.9)^2}{12.9} + \frac{(9-9.9)^2}{9.9} + \frac{(6-6.4)^2}{6.4} + \frac{(5-5.5)^2}{5.5} + \frac{(9-13.5)^2}{13.5} = 5.624 \text{ (df} = 6)$$
4. H_0: Whether getting an infection is independent of type of treatment.
   H_1: Getting an infection depends on type of treatment.
   Test statistic: $\chi^2 = 0.773$; P-value $= 0.856$ (Table: P-value > 0.10); Critical value: $\chi^2 = 11.345$; $\text{df} = (2-1)(4-1) = 3$; Fail to reject H_0. There is not sufficient evidence to warrant rejection of the claim that getting an infection is independent of the treatment. The atorvastatin (Lipitor) treatment does not appear to have an effect on infections.
   $$\chi^2 = \frac{(27-27.08)^2}{27.08} + \cdots + \frac{(7-9.43)^2}{9.43} + \frac{(234-242.92)^2}{242.92} + \cdots + \frac{(87-84.57)^2}{84.57} = 0.773$$

5. H_0: $p_{\text{Jan}} = p_{\text{Feb}} = p_{\text{Mar}} = p_{\text{Apr}} = p_{\text{May}} = p_{\text{Jun}} = p_{\text{Jul}} = p_{\text{Aug}} = p_{\text{Sep}} = p_{\text{Oct}} = p_{\text{Nov}} = p_{\text{Dec}} = 1/12$;
   H_1: At least one of the proportions is not equal to the given claimed value.

   Test statistic: $\chi^2 = 269.147$, P-value = 0.0000 (Table: P-value < 0.005); Critical value: $\chi^2 = 24.725$; Reject H_0. There is sufficient evidence to warrant rejection of the claim that weather related deaths occur in the different months with the same frequency. The months of May, June, and July appear to have disproportionately more weather-related deaths, and that is probably due to the fact that vacations and outdoor activities are much greater during those months.

$$\chi^2 = \frac{(28-37.5)^2}{450/12} + \frac{(17-37.5)^2}{450/12} + \cdots + \frac{(26-37.5)^2}{450/12} + \frac{(25-37.5)^2}{450/12} = 269.147 \text{ (df} = 11)$$

Cumulative Review Exercises

1. $H_0: p = 0.5$; $H_1: p \neq 0.5$; Test statistic: $z = \dfrac{\frac{320}{450} - 0.5}{\sqrt{\frac{(0.5)(0.5)}{450}}} = 8.96$;

   $P\text{-value} = 2 \cdot P(z > 8.96) = 0.0000$ (Table: 0.0002); Critical values: $z = \pm 1.96$;

   Reject H_0. There is sufficient evidence to warrant rejection of the claim that among those who die in weather-related deaths, the percentage of males is equal to 50%. One possible explanation is that more men participate in some outdoor activities, such as golf, fishing, and boating.

2. a. There is a possibility that the results were affected because the sponsor of the survey produces chocolate and therefore has an interest in the results.

   b. $0.85 \cdot 1708 = 1452$

3. 99% CI: $\hat{p} \pm z_{\alpha/2}\sqrt{\dfrac{\hat{p}\hat{q}}{n}} = 0.85 \pm 2.756\sqrt{\dfrac{(0.85)(0.15)}{1708}} \Rightarrow 0.828 < p < 0.872$, or $82.8\% < p < 87.2\%$; We have 99% confidence that the limits of 82.8% and 87.2% contain the value of the true percentage of the population of women saying that chocolate makes them happier.

4. $H_0: p = 0.80$; $H_1: p > 0.80$; Test statistic: $z = \dfrac{\frac{1452}{1708} - 0.80}{\sqrt{\frac{(0.80)(0.20)}{1708}}} = 5.18$;

   $P\text{-value} = P(z > 5.18) = 0.0000$ (Table: 0.0001); Critical value: $z = 2.33$;

   Reject H_0. There is sufficient evidence to support the claim that when asked, more than 80% of women say that chocolate makes them happier.

5. H_0: Whether money was spent is independent of form of 100 Yuan.
   H_1: Whether money was spent depends on form of 100 Yuan.

   Test statistic: $\chi^2 = 3.409$; $P\text{-value} = 0.0648$ (Table: P-value > 0.05); Critical value: $\chi^2 = 3.841$; df $= (2-1)(2-1) = 1$; Fail to reject H_0. There is not sufficient evidence to warrant rejection of the claim that the form of the 100-yuan gift is independent of whether the money was spent. There is not sufficient evidence to support the claim of a denomination effect. Women in China do not appear to be affected by whether 100 Yuan are in the form of a single bill or several smaller bills.

$$\chi^2 = \frac{(60-64)^2}{64} + \frac{(15-11)^2}{11} + \frac{(68-64)^2}{64} + \frac{(7-11)^2}{11} = 3.409$$

6. a. $\frac{60+68}{150}=\frac{128}{150}=\frac{64}{75}=0.853$

   b. $\frac{60+68}{150}+\frac{60+15}{150}-\frac{60}{150}=\frac{143}{150}=0.953$

   c. $\frac{128}{150}\cdot\frac{127}{149}=\frac{8128}{11{,}175}=0.727,\ \frac{128}{150}\cdot\frac{128}{150}=0.728$

7. $r=-0.283$; P-value $=0.539$; Critical values $(\alpha=0.05)$: $r=\pm 0.754$; There is not sufficient evidence to support the claim of a linear correlation between the repair costs from full-front crashes and full-rear crashes.

8. a. The z score for the bottom 5% is -1.645, which correspond a forward grip reach of $-1.645\cdot 34+686$ $=630$ mm.

   b. $z_{x=650}=\frac{650-686}{34}=-1.06$; which has a probability of 0.1448, or 14.48% (Table: 14.46%) to the left. That percentage is too high, because too many women would not be accommodated.

   c. $z_{x=680}=\frac{680-686}{34/\sqrt{16}}=-0.71$; which has a probability of $1-0.2401=0.7599$ (Table: 0.7611) to the right. Groups of 16 women do not occupy a driver's seat or cockpit; because individual women occupy the driver's seat/cockpit, this result has no effect on the design.

Chapter 12: Analysis of Variance

Section 12-1: One-Way ANOVA

1. a. The arrival delay times are categorized according to the one characteristic of the flight number.
   b. The terminology of *analysis of variance* refers to the method used to test for equality of the three population means. That method is based on two different estimates of a common population variance.
3. The test statistic is $F = 1.334$, and the F distribution applies.
5. Test statistic: $F = 0.39$; P-value: 0.677; Fail to reject H_0: $\mu_1 = \mu_2 = \mu_3$. There is not sufficient evidence to warrant rejection of the claim that the three categories of blood lead level have the same mean verbal IQ score. Exposure to lead does not appear to have an effect on verbal IQ scores.
7. Test statistic: $F = 5.5963$; P-value: 0.0045; Reject H_0: $\mu_1 = \mu_2 = \mu_3$. There is sufficient evidence to warrant rejection of the claim that the three samples are from populations with the same mean. It appears that at least one of the mean service times is different from the others.
9. Test statistic: $F = 7.9338$; P-value: 0.0005; Reject H_0: $\mu_1 = \mu_2 = \mu_3$. There is sufficient evidence to warrant rejection of the claim that females from the three age brackets have the same mean pulse rate. It appears that pulse rates of females are affected by age bracket.
11. Test statistic: $F = 27.2488$; P-value: 0.000; Reject H_0: $\mu_1 = \mu_2 = \mu_3$. There is sufficient evidence to warrant rejection of the claim that the three different miles have the same mean time. These data suggest that the third mile appears to take longer, and a reasonable explanation is that the third lap has a hill.

EXCEL
ANOVA

Source of Variation	*SS*	*df*	*MS*	*F*	*P-value*	*F crit*
Between Groups	0.103444	2	0.051722	27.24878	3.45E-05	3.885294
Within Groups	0.022778	12	0.001898			
Total	0.126222	14				

13. Test statistic: $F = 2.3163$; P-value: 0.123; Fail to reject H_0: $\mu_1 = \mu_2 = \mu_3$. There is not sufficient evidence to warrant rejection of the claim that the three different flights have the same mean departure delay time. The departure delay times from Flight 1 have very little variation, and departures of Flight 1 appear to be on time or slightly early. Departure delay times from Flight 21 appear to have considerable variation. With variances of 2.5 min^2, 709.8 min^2, and 2525.4 min^2. the ANOVA requirement of the same variance appears to be violated even for this loose requirement.

EXCEL
ANOVA

Source of Variation	*SS*	*df*	*MS*	*F*	*P-value*	*F crit*
Between Groups	5028.074	2	2514.037	2.63723	0.092187	3.402826
Within Groups	22878.89	24	953.287			
Total	27906.96	26				

15. Test statistic: $F = 28.1666$; P-value: 0.000; Reject H_0: $\mu_1 = \mu_2 = \mu_3$. There is sufficient evidence to warrant rejection of the claim that the three different types of Chips Ahoy cookies have the same mean number of chocolate chips. The reduced fat cookies have a mean of 19.6 chocolate chips, which is slightly more than the mean of 19.1 chocolate chips for the chewy cookies, so the reduced fat does not appear to be the result of including fewer chocolate chips. Perhaps the fat content in the chocolate chips is different and/or the fat content in the cookie material is different.

15. (continued)

EXCEL

ANOVA

Source of Variation	*SS*	*df*	*MS*	*F*	*P-value*	*F crit*
Between Groups	542.772321	2	271.386161	28.1666001	1.3776E-10	3.07959586
Within Groups	1050.21875	109	9.6350344			
Total	1592.99107	111				

17. The Tukey test results show different *P*-values, but they are not dramatically different. The Tukey results suggest the same conclusions as the Bonferroni test.

Section 12-2: Two-Way ANOVA

1. The pulse rates are categorized using two different factors of (1) age bracket and (2) gender.

3. a. An interaction between two factors or variables occurs if the effect of one of the factors changes for different categories of the other factor.
   b. If there is an interaction effect, we should not proceed with individual tests for effects from the row factor and column factor. If there is an interaction, we should not consider the effects of one factor without considering the effects of the other factor.
   c. Because the lines are far from parallel, the two genders have very different effects for the different age brackets, so there does appear to be an interaction between gender and age bracket.

5. For interaction, the test statistic is $F = 9.58$ and the *P*-value is 0.0003, so there is sufficient evidence to warrant rejection of the null hypothesis of no interaction effect. Because there appears to be an interaction between age bracket and gender, we should not proceed with a test for an effect from age bracket and a test for an effect from gender. It appears an interaction between age bracket and gender has an effect on pulse rates. (Remember, these results are based on fabricated data used in one of the cells, so this conclusion does not necessarily correspond to real data.)

7. For interaction, the test statistic is $F = 1.7970$ and the *P*-value is 0.1756, so there is not sufficient evidence to conclude that there is an interaction effect. For the row variable of age bracket, the test statistic is $F = 2.0403$ and the *P*-value is 0.1399, so there is not sufficient evidence to conclude that age bracket has an effect on height. For the column variable of gender, the test statistic is $F = 43.4607$ and the *P*-value is less than 0.0001, so there is sufficient evidence to support the claim that gender has an effect on height.

9. For interaction, the test statistic is $F = 1.1653$ and the *P*-value is 0.3289, so there is no significant interaction effect. For gender, the test statistic is 1.6864 and the *P*-value is 0.2064, so there is no significant effect from gender. For age, the test statistic is $F = 5.0998$ and the *P*-value is 0.0143, so there is a significant effect from age.

EXCEL

ANOVA

Source of Variation	*SS*	*df*	*MS*	*F*	*P-value*	*F crit*
Rows	15225413	1	15225413	1.686377	0.206419	4.259677
Columns	92086979	2	46043490	5.099807	0.014265	3.402826
Interaction	21042069	2	10521034	1.165317	0.328851	3.402826
Total	3.45E+08	29				

11. a. Test statistics and *P*-values do not change.
    b. Test statistics and *P*-values do not change.
    c. Test statistics and *P*-values do not change.
    d. An outlier can dramatically affect and change test statistics and *P*-values.

Quick Quiz

1. H_0: $\mu_1 = \mu_2 = \mu_3$; Because the displayed P-value of 0.000 is small, reject H_0. There is sufficient evidence to warrant rejection of the claim that the four samples have the same mean weight.
2. No, it appears that mean weights of Diet Coke and Diet Pepsi are lower than the mean weights of regular Coke and regular Pepsi, but the method of analysis of variance does not justify a conclusion that any particular means are significantly different from the others.
3. right-tailed.
4. Test statistic: $F = 503.06$; Larger test statistics result in smaller P-values.
5. The four samples are categorized using only one factor: the type of cola (regular Coke, Diet Coke, regular Pepsi, Diet Pepsi).
6. One-way analysis of variance is used to test a null hypothesis that three or more samples are from populations with equal means.
7. With one-way analysis of variance, data from the different samples are categorized using only one factor, but with two-way analysis of variance, the sample data are categorized into different cells determined by two different factors.
8. Fail to reject the null hypothesis of no interaction. There does not appear to be an effect due to an interaction between sex and major.
9. There is not sufficient evidence to support a claim that the length estimates are affected by the sex of the subject.
10. There is not sufficient evidence to support a claim that the length estimates are affected by the subject's major.

Review Exercises

1. Test statistic: $F = 2.7347$; P-value: 0.0829; Fail to reject H_0: $\mu_1 = \mu_2 = \mu_3$. There is not sufficient evidence to warrant rejection of the claim that males in the different age brackets give attribute ratings with the same mean. Age does not appear to be a factor in the male attribute ratings.
2. Test statistic: $F = 9.4695$; P-value: 0.0006; Reject H_0: $\mu_1 = \mu_2 = \mu_3$ There is sufficient evidence to warrant rejection of the claim that the three books have the same mean Flesch Reading Ease score. The data suggest that the books appear to have mean scores that are not all the same, so the authors do not appear to have the same level of readability.

   EXCEL
   ANOVA

Source of Variation	*SS*	*df*	*MS*	*F*	*P-value*	*F crit*
Between Groups	1338.002	2	669.0011	9.469487	0.000562	3.284918
Within Groups	2331.387	33	70.64808			
Total	3669.389	35				

3. For interaction, the test statistic is $F = 1.7171$ and the P-value is 0.1940, so there is not sufficient evidence to warrant rejection of no interaction effect. There does not appear to be an interaction between femur and car size. For the row variable of femur, the test statistic is $F = 1.3896$ and the P-value is 0.2462, so there is not sufficient evidence to conclude that whether the femur is right or left has an effect on load. For the column variable of car size, the test statistic is $F = 2.2296$ and the P-value is 0.1222, so there is not sufficient evidence to warrant rejection of the claim of no effect from car size. It appears that the crash test loads are not affected by an interaction between femur and car size, they are not affected by femur, and they are not affected by car size.

4. For interaction, the test statistic is $F = 0.4784$ and the P-value is 0.7513, so there is not sufficient evidence to conclude that there is an interaction effect. For the row variable of age bracket of females, the test statistic is $F = 0.3149$ and the P-value is 0.7318, so there is not sufficient evidence to conclude that the age bracket of females has an effect on the ratings. For the column variable of age bracket of males, the test statistic is $F = 1.1939$ and the P-value is 0.3148, so there is not sufficient evidence to conclude that the age bracket of males has an effect on the ratings.

EXCEL

ANOVA

Source of Variation	*SS*	*df*	*MS*	*F*	*P-value*	*F crit*
Rows	30.57778	2	15.28889	0.314946	0.731819	3.259446
Columns	115.9111	2	57.95556	1.193866	0.314761	3.259446
Interaction	92.88889	4	23.22222	0.47837	0.751347	2.633532
Total	1986.978	44				

Cumulative Review Exercises

1. a. Flight 3: $\bar{x} = 2.0$ min; Flight 19: $\bar{x} = 9.9$ min; Flight 21: $\bar{x} = 33.4$ min
   b. Flight 3: $s = 10.6$ min; Flight 19: $s = 26.6$ min; Flight 21: $s = 50.3$ min
   c. Flight 3: $s^2 = 112.0\text{ min}^2$; Flight 19: $s^2 = 709.8\text{ min}^2$; Flight 21: $s^2 = 2524.4\text{ min}^2$
   d. The departure delay time of 142 min is an outlier.
   e. ratio
2. H_0: $\mu_1 = \mu_2$; H_1: $\mu_1 \neq \mu_2$; population$_1$ = Flight 3, population$_2$ = Flight 21;
   Test statistic: $t = -1.728$; P-value $= 0.1241$ (Table: P-value > 0.10); Critical values ($\alpha = 0.05$): $t = \pm 2.326$ (Table: $t = \pm 2.365$); Fail to reject H_0. There is not sufficient evidence to support the claim that there is a difference between the departure delay times for the two flights.
   $$t = \frac{(\bar{x}_1 - \bar{x}_2) - (\mu_1 - \mu_2)}{\sqrt{\frac{s_1^2}{n_1} + \frac{s_2^2}{n_2}}} = \frac{(2.0 - 33.4) - 0}{\sqrt{\frac{10.6^2}{8} + \frac{50.3^2}{8}}} = -1.728 \text{ (df} = 7)$$
3. Because the pattern of the points is far from a straight-line pattern, the departure delay times for Flight 19 do not appear to be from a population with a normal distribution.
4. The data appear to fit the loose definition of a normally distribution.
   95% CI: $\bar{x} \pm t_{\alpha/2} \frac{s}{\sqrt{n}} = 2.0 \pm 2.365 \cdot \frac{26.6}{\sqrt{8}} \Rightarrow -6.8\text{ min} < \mu < 10.8\text{ min}$; We have 95% confidence that the limits of -6.8 min and 10.8 min contain the value of the population mean for all Flight 3 departure delays.
5. a. H_0: $\mu_1 = \mu_2 = \mu_3$
   b. Because the P-value of 0.1729 is greater than the significance level of 0.05, fail to reject the null hypothesis of equal means. There is not sufficient evidence to warrant rejection of the claim that the three means are equal. The three populations do not appear to have means that are significantly different.

6. a. $z_{x=5.600} = \frac{5.600 - 5.670}{0.062} = -1.13$ and $z_{x=5.700} = \frac{5.700 - 5.670}{0.062} = 0.48$ which have a probability of $0.6844 - 0.1292 = 0.5552$ (Tech: 0.5552) between them.

   b. $z_{x=5.675} = \frac{5.675 - 5.670}{0.062/\sqrt{25}} = 0.40$; which has a probability of $1 - 0.6554 = 0.3446$ (Tech: 0.3434) to the right.

   c. Half the quarters will weigh less than the mean of 5.670 g, so the probability that eight quarters selected randomly all weighing less than 5.670 g is $(1/2)^8 = 1/256$, or about 0.00391.

   d. The z score for the bottom 10% is –1.28, which correspond to a weight of $-1.28 \cdot 0.062 + 5.760 = 5.591\text{ g}$.

7. a. $0.20(1000) = 200$

   b. 95% CI: $\hat{p} \pm z_{\alpha/2}\sqrt{\frac{\hat{p}\hat{q}}{n}} = 0.20 \pm 1.96\sqrt{\frac{(0.20)(0.80)}{1000}} \Rightarrow 0.175 < p < 0.225$

   c. Yes, the confidence interval shows us that we have 95% confidence that the true population proportion is contained within the limits of 0.175 and 0.225, and $1/4$ is not included within that range.

8. a. Because the vertical scale begins at 15 instead of 0, the graph is deceptive by exaggerating the differences among the frequencies.

   b. No, a normal distribution is approximately bell-shaped, but the given histogram is far from being bell-shaped. Because the digits are supposed to be equally likely, the histogram should be flat with all bars having approximately the same height.

   c. The frequencies are 19, 21, 22, 21, 18, 23, 16, 16, 22, and 22.

   d. H_0: $p_0 = p_1 = p_2 = p_3 = p_4 = p_5 = p_6 = p_7 = p_8 = p_9 = 0.10$;

   H_1: At least one of the proportions is not equal to the given claimed value.

   Test statistic: $\chi^2 = 3.000$; P-value = 0.964 (Table: P-value > 0.95); Critical value ($\alpha = 0.05$): $\chi^2 = 16.919$; Fail to reject H_0. There is not sufficient evidence to warrant rejection of the claim that the digits are selected from a population in which the digits are all equally likely. There does not appear to be a problem with the lottery.

$$\chi^2 = \frac{(19-20)^2}{0.10 \cdot 200} + \frac{(21-20)^2}{0.10 \cdot 200} + \cdots + \frac{(22-20)^2}{0.10 \cdot 200} + \frac{(22-20)^2}{0.10 \cdot 200} = 3.000 \text{ (df = 9)}$$

Chapter 13: Nonparametric Tests

13-2: Sign Test

1. The only requirement for the matched pairs is that they constitute a simple random sample. There is no requirement of a normal distribution or any other specific distribution. The sign test is "distribution free" in the sense that it does not require a normal distribution or any other specific distribution.

3. H_0: There is no difference between the populations of September weights and the matching April weights.

   H_1: There is a difference between the populations of September weights and the matching April weights.

   The sample data do not contradict H_1 because the numbers of positive signs (2) and negative signs (7) are not exactly the same.

5. The test statistic of $x = 3$ is not less than or equal to the critical value of 1 (from Table A-7). There is not sufficient evidence to warrant rejection of the claim of no difference. There is not sufficient evidence to support the claim that there is a difference between female attribute ratings and male attribute ratings.

7. The test statistic of $z = \frac{(82+0.5)-\frac{187}{2}}{\sqrt{187}/2} = -1.61$ results in a P-value of 0.1074, and it does not fall in the critical region bounded by $z = -1.96$ and 1.96. There is not sufficient evidence to warrant rejection of the claim of no difference. There is not sufficient evidence to support the claim that there is a difference between female attribute ratings and male attribute ratings.

9. The test statistic of $z = \frac{(401+0.5)-\frac{882}{2}}{\sqrt{882}/2} = -2.66$ results in a P-value of 0.0078, and it is in the critical region bounded by $z = -2.575$ and 2.575. There is sufficient evidence to warrant rejection of the claim that there is no difference between the proportions of those opposed and those in favor.

11. The test statistic of $z = \frac{(426+0.5)-\frac{860}{2}}{\sqrt{860}/2} = -0.24$ results in a P-value of 0.8103, and it is not in the critical region bounded by $z = -1.96$ and 1.96. There is not sufficient evidence to reject the claim that boys and girls are equally likely.

13. The test statistic of $z = \frac{(116+0.5)-\frac{598}{2}}{\sqrt{598}/2} = -14.93$ results in a P-value of 0.0000, and it is in the critical region bounded by $z = -2.575$ and 2.575. There is sufficient evidence to warrant rejection of the claim that the median is equal to 2.00.

15. The test statistic of $z = \frac{(12+0.5)-\frac{40}{2}}{\sqrt{40}/2} = -2.37$ results in a P-value of 0.0178 and it is not in the critical region bounded by $z = -2.575$ and 2.575. There is not sufficient evidence to warrant rejection of the claim that the median is equal to 5.670 g. The quarters appear to be minted according to specifications.

17. Second approach: The test statistic of $z = \frac{(30+0.5)-\frac{105}{2}}{\sqrt{105}/2} = -4.29$ is in the critical region bounded by $z = -1.645$, so the conclusions are the same as in Example 4.

    Third approach: The test statistic of $z = \frac{(38+0.5)-\frac{106}{2}}{\sqrt{106}/2} = -2.82$ is in the critical region bounded by $z = -1.645$, so the conclusions are the same as in Example 4. The different approaches can lead to very different results; as seen in the test statistics of –4.21, –4.29, and –2.82. The conclusions are the same in this case, but they could be different in other cases.

13-3: Wilcoxon Signed-Ranks Test for Matched Pairs

1. a. The only requirements are that the matched pairs be a simple random sample and the population of differences be approximately symmetric.
   b. There is no requirement of a normal distribution or any other specific distribution.
   c. The Wilcoxon signed-ranks test is "distribution free" in the sense that it does not require a normal distribution or any other specific distribution.

3. The sign test uses only the signs of the differences, but the Wilcoxon signed-ranks test uses ranks that are affected by the magnitudes of the differences.

5. Test statistic: $T = 16.5$; Critical value: $T = 6$; Fail to reject the null hypothesis that the population of differences has a median of 0. There is not sufficient evidence to support the claim that there is a difference between female attribute ratings and male attribute ratings.

7. Convert $T = 8323.5$ to the test statistic $z = -0.63$. P-value: 0.5287; Critical values: $z = \pm 1.96$; There is not sufficient evidence to warrant rejection of the claim of no difference. There is not sufficient evidence to support the claim that there is a difference between female attribute ratings and male attribute ratings.

$$z = \frac{T - \frac{n(n+1)}{4}}{\sqrt{\frac{n(n+1)(2n+1)}{24}}} = \frac{8323.5 - \frac{187(187+1)}{4}}{\sqrt{\frac{187(187+1)(2\cdot 187+1)}{24}}} = -0.63$$

9. Convert $T = 18{,}014$ to the test statistic $z = -16.92$. P-value: 0.0000. Critical values: $z = \pm 2.575$; There is sufficient evidence to warrant rejection of the claim that the median is equal to 2.00.

$$z = \frac{T - \frac{n(n+1)}{4}}{\sqrt{\frac{n(n+1)(2n+1)}{24}}} = \frac{18{,}014 - \frac{598(598+1)}{4}}{\sqrt{\frac{598(598+1)(2\cdot 598+1)}{24}}} = -16.92$$

11. Convert $T = 196$ to the test statistic $z = -2.88$. P-value: 0.0040. Critical values: $z = \pm 2.575$; There is sufficient evidence to warrant rejection of the claim that the median is equal to 5.670 g. The quarters do not appear to be minted according to specifications.

$$z = \frac{T - \frac{n(n+1)}{4}}{\sqrt{\frac{n(n+1)(2n+1)}{24}}} = \frac{196 - \frac{40(40+1)}{4}}{\sqrt{\frac{40(40+1)(2\cdot 40+1)}{24}}} = -1.42$$

13. a. Smallest: $T = 0$; Largest: $T = 1 + 2 + 3 + \cdots + 248 + 249 + 250 = \frac{250(250+1)}{2} = 31{,}375$
    b. $\frac{31{,}375}{2} = 15687.5$
    c. $31{,}375 - 1234 = 30{,}141$
    d. $\frac{n(n+1)}{2} - k$

13-4: Wilcoxon Rank-Sum Test for Two Independent Samples

1. Yes, the two samples are independent. The evaluations of female professors and male professors are not matched in any way. The samples are simple random samples. Each sample has more than 10 values.

3. H_0: Evaluations of female professors and male professors have the same median. There are three different possible alternative hypotheses: H_1: Evaluations of female professors and male professors have different medians. H_1: Evaluations of female professors have a median greater than the median of male professor evaluations. H_1: Evaluations of female professors have a median less than the median of male professor evaluations.

5. $R_1 = 163$; $R_2 = 188$; $\mu_R = 189$; $\sigma_R = 19.4422$; Test statistic: $z = -1.34$; P-value $= 0.1802$; Critical values: $z = \pm 1.96$; Fail to reject the null hypothesis that the populations have the same median. There is not sufficient evidence to warrant rejection of the claim that evaluation ratings of female professors have the same median as evaluation ratings of male professors.

$$\mu_R = \frac{n_1(n_1 + n_2 + 1)}{2} = \frac{14(14 + 12 + 1)}{2} = 189$$

$$\sigma_R = \sqrt{\frac{n_1 n_2 (n_1 + n_2 + 1)}{12}} = \sqrt{\frac{14 \cdot 12(14 + 12 + 1)}{12}} = 19.4422$$

$$z = \frac{Z - \mu_R}{\sigma_R} = \frac{163 - 189}{19.4422} = -1.34$$

7. $R_1 = 253.5$; $R_2 = 124.5$; $\mu_R = 182$; $\sigma_R = 20.607$; Test statistic: $z = 3.47$; P-value $= 0.0005$; Critical values: $z = \pm 1.96$; Reject the null hypothesis that the populations have the same median. There is sufficient evidence to reject the claim that for those treated with 20 mg of Lipitor and those treated with 80 mg of Lipitor, changes in LDL cholesterol have the same median. It appears that the dosage amount does have an effect on the change in LDL cholesterol.

$$\mu_R = \frac{n_1(n_1 + n_2 + 1)}{2} = \frac{13(13 + 14 + 1)}{2} = 182$$

$$\sigma_R = \sqrt{\frac{n_1 n_2 (n_1 + n_2 + 1)}{12}} = \sqrt{\frac{13 \cdot 14(13 + 14 + 1)}{12}} = 20.607$$

$$z = \frac{Z - \mu_R}{\sigma_R} = \frac{253.5 - 182}{20.607} = 3.47$$

9. $R_1 = 1615.5$; $R_2 = 2755.5$; $\mu_R = 1880$; $\sigma_R = 128.8669$; Test statistic: $z = -2.05$; P-value $= 0.0404$; Critical values: $z = \pm 1.96$; Reject the null hypothesis that the populations have the same median. There is sufficient evidence to warrant rejection of the claim that evaluation ratings of female professors have the same median as evaluation ratings of male professors.

$$\mu_R = \frac{n_1(n_1 + n_2 + 1)}{2} = \frac{40(40 + 53 + 1)}{2} = 1880$$

$$\sigma_R = \sqrt{\frac{n_1 n_2 (n_1 + n_2 + 1)}{12}} = \sqrt{\frac{40 \cdot 53(40 + 53 + 1)}{12}} = 128.8669$$

$$z = \frac{Z - \mu_R}{\sigma_R} = \frac{1615.5 - 1880}{128.8669} = -2.05$$

11. $R_1 = 501$; $R_2 = 445$; $\mu_R = 484$; $\sigma_R = 41.15823$; Test statistic: $z = 0.41$; P-value $= 0.3409$; Critical value: $z = 1.645$; Fail to reject the null hypothesis that the populations have the same median. There is not sufficient evidence to support the claim that subjects with medium lead levels have a higher median of the full IQ scores than subjects with high lead levels. Based on these data, it does not appear that lead level affects full IQ scores.

$$\mu_R = \frac{n_1(n_1 + n_2 + 1)}{2} = \frac{22(22 + 21 + 1)}{2} = 484$$

$$\sigma_R = \sqrt{\frac{n_1 n_2 (n_1 + n_2 + 1)}{12}} = \sqrt{\frac{22 \cdot 21(22 + 21 + 1)}{12}} = 41.158$$

$$z = \frac{Z - \mu_R}{\sigma_R} = \frac{501 - 484}{41.15823} = 0.41$$

13. The test statistic is the same value with opposite sign.

$$U = n_1 n_2 + \frac{n_1(n_1+1)}{2} - R = 12 \cdot 15 + \frac{12(12+1)}{2} - 159.5 = 98.5$$

$$z = \frac{U - \frac{n_1 n_2}{2}}{\sqrt{\frac{n_1 n_2 (n_1 + n_2 + 1)}{12}}} = \frac{98.5 - \frac{12 \cdot 15}{2}}{\sqrt{\frac{12 \cdot 15 \cdot (12+15+1)}{12}}} = 0.41$$

13-5: Kruskal-Wallis Test for Three or More Samples

1. $R_1 = 164.5;\ R_2 = 150;\ R_3 = 150.5$

Age 20-22	Age 23-26	Age 27-29
38.0 (21.0)	39.0 (22.5)	36.0 (17.0)
42.0 (25.5)	31.0 (7.5)	42.0 (25.5)
30.0 (6.0)	36.0 (17.0)	35.5 (14.0)
39.0 (22.5)	35.0 (13.0)	27.0 (4.0)
47.0 (29.5)	41.0 (24.0)	37.0 (20.0)
43.0 (27.0)	45.0 (28.0)	34.0 (12.0)
33.0 (11.0)	36.0 (17.0)	22.0 (2.0)
31.0 (7.5)	23.0 (3.0)	47.0 (29.5)
32.0 (9.5)	36.0 (17.0)	36.0 (17.0)
28.0 (5.0)	20.0 (1.0)	32.0 (9.5)

(Ranks for each value shown in parentheses.)

3. $n_1 = 10,\ n_2 = 10,\ n_3 = 10$, and $N = 10 + 10 + 10 = 30$

5. Test statistic: $H = 0.1748$; Critical value: $\chi^2 = 5.991$; (Tech: *P*-value $= 0.916$); Fail to reject the null hypothesis of equal medians. There is not sufficient evidence to warrant rejection of the claim that females from the different age brackets give attribute ratings with the same median.

$$H = \frac{12}{N(N+1)}\left(\frac{R_1^2}{n_1} + \frac{R_2^2}{n_2} + \frac{R_3^2}{n_3}\right) - 3(N+1) = \frac{12}{30(30+1)}\left(\frac{164.5^2}{10} + \frac{150^2}{10} + \frac{150.5^2}{10}\right) - 3(30+1)$$
$$= 0.1748$$

7. Test statistic: $H = 4.9054$; Critical value: $\chi^2 = 5.991$; (Tech: *P*-value $= 0.086$); Fail to reject the null hypothesis of equal medians. The data do not suggest that larger cars are safer.

$$H = \frac{12}{N(N+1)}\left(\frac{R_1^2}{n_1} + \frac{R_2^2}{n_2} + \frac{R_3^2}{n_3}\right) - 3(N+1) = \frac{12}{21(21+1)}\left(\frac{86^2}{7} + \frac{97^2}{7} + \frac{48^2}{7}\right) - 3(21+1)$$
$$= 4.9054$$

9. Test statistic: $H = 11.4704$; Critical value: $\chi^2 = 5.991$; (Tech: *P*-value $= 0.003$); Reject the null hypothesis of equal medians. It appears that the three restaurants have dinner drive-through service times with different medians.

$$H = \frac{12}{N(N+1)}\left(\frac{R_1^2}{n_1} + \frac{R_2^2}{n_2} + \frac{R_3^2}{n_3}\right) - 3(N+1) = \frac{12}{150(150+1)}\left(\frac{4580.5^2}{50} + \frac{3606^2}{50} + \frac{3138.5^2}{50}\right) - 3(150+1)$$
$$= 11.4704$$

11. Test statistic: $H = 2.5999$; Critical value: $\chi^2 = 7.815$; (Tech: P-value = 0.458); Fail to reject the null hypothesis of equal medians. It appears that the four hospitals have birth weights with the same median.

$$H = \frac{12}{N(N+1)}\left(\frac{R_1^2}{n_1}+\frac{R_2^2}{n_2}+\frac{R_3^2}{n_3}+\frac{R_4^2}{n4}\right)-3(N+1)$$

$$= \frac{12}{400(400+1)}\left(\frac{20{,}189^2}{100}+\frac{21{,}262.5^2}{100}+\frac{20{,}106.5^2}{100}+\frac{18{,}642^2}{100}\right)-3(400+1) = 2.5999$$

13. The values of t are 2, 2, 2, 2, and 4. (See table below) The values of T are 6, 6, 6, 6, and 60 and $\Sigma T = 84$. Using $\Sigma T = 84$ and $N = 8+6+5 = 19$, the corrected value of H is $0.694\Big/\left(1-\frac{84}{19^3-19}\right) = 0.703$, which is not substantially different from the value of 0.694 found in Example 1. In this case, the large numbers of ties do not appear to have a dramatic effect on the test statistic H.

Lead Level	Rank	t	$t^3 - t$
85	1.5	2	6
90	4.5	2	6
97	6.5	4	60
100	17.0	2	6
107	28.0	2	6
		SUM	84

13-6: Rank Correlation

1. The methods of Section 10-2 should not be used for predictions. The regression equation is based on a linear correlation between the two variables, but the methods of this section do not require a linear relationship. The methods of this section could suggest that there is a correlation with paired data associated by some nonlinear relationship, so the regression equation would not be a suitable model for making predictions.

3. r represents the linear correlation coefficient computed from sample paired data; r represents the parameter of the linear correlation coefficient computed from a population of paired data; r_s denotes the rank correlation coefficient computed from sample paired data; ρ_s represents the rank correlation coefficient computed from a population of paired data. The subscript s is used so that the rank correlation coefficient can be distinguished from the linear correlation coefficient r. The subscript does not represent the standard deviation s. It is used in recognition of Charles Spearman, who introduced the rank correlation method.

5. $r_s = 1.000$; Critical values are –0.886 and 0.886. Reject the null hypothesis of $\rho_s = 0$. There is sufficient evidence to support a claim of a correlation between distance and time.

7. $r_s = 0.888$; Critical values: –0.618 and 0.618. Reject the null hypothesis of $\rho_s = 0$. There is sufficient evidence to support the claim of a correlation between chocolate consumption and the rate of Nobel Laureates. It does not make sense to think that there is a cause/effect relationship, so the correlation could be the result of a coincidence or other factors that affect the variables the same way.

9. $r_s = 1.000$; Critical values: –0.700 and 0.700. Reject the null hypothesis of $\rho_s = 0$. There is sufficient evidence to support the claim of a correlation between the cost of a slice of pizza and the subway fare.

11. $r_s = 1.000$; Critical values: -0.886, 0.886. Reject the null hypothesis of $\rho_s = 0$. There is sufficient evidence to conclude that there is a correlation between overhead widths of seals from photographs and the weights of the seals.

13. $r_s = 0.902$; Critical values: $r_s = \frac{\pm z}{\sqrt{n-1}} = \frac{\pm 1.96}{\sqrt{23-1}} = \pm 0.418$; Reject the null hypothesis of $\rho_s = 0$. There is sufficient evidence to support the claim of a correlation between chocolate consumption and the rate of Nobel Laureates. It does not make sense to think that there is a cause/effect relationship, so the correlation could be the result of a coincidence or other factors that affect the variables the same way.

15. $r_s = 0.360$; Critical values: $r_s = \frac{\pm z}{\sqrt{n-1}} = \frac{\pm 1.96}{\sqrt{153-1}} = \pm 0.159$; Reject the null hypothesis of $\rho_s = 0$. There is sufficient evidence to conclude that there is a correlation between the systolic and diastolic blood pressure levels in males.

17. $r_s = \pm\sqrt{\frac{1.975799^2}{1.975799^2 + 153 - 2}} = \pm\sqrt{\frac{1.978^2}{1.978^2 + 153 - 2}} = \pm 0.159$; (Use either $t = 1.975799$ from technology or use interpolation in Table A-3 with 151 degrees of freedom, so the critical value of t is approximately halfway between 1.984 and 1.972, which is 1.978.) The critical values are the same as those found by using Formula 13-1.

13-7: Runs Test for Randomness

1. No, the runs test can be used to determine whether the sequence of political parties is not random, but the runs test does not show whether the proportion of Republicans is significantly greater than the proportion of Democrats.

3. The critical values are 8 and 19. Because $G = 16$ is not less than or equal to 8 nor is $G = 16$ greater than or equal to 18, fail to reject randomness. It appears that the sequence of political parties is random.

5. $\bar{x} = 157.7$ fatalities; $n_1 = 11$; $n_2 = 9$; $G = 12$; critical values: 6, 16; Fail to reject randomness. There is not sufficient evidence to warrant rejection of the claim that there is randomness above and below the mean. There does not appear to be a trend.

7. $n_1 = 20$; $n_2 = 10$; $G = 16$; critical values: 9, 20; Fail to reject randomness. There is not sufficient evidence to reject the claim that the dates before and after July 1 are randomly selected.

9. $n_1 = 26$; $n_2 = 23$; $G = 20$; $\mu_G = 25.40816$; $\sigma_G = 3.450091$; Test statistic: $z = -1.57$; P-value $= 0.1164$; Critical values: $z = \pm 1.96$; Fail to reject randomness. There is not sufficient evidence to reject randomness. The runs test does not test for disproportionately more occurrences of one of the two categories, so the runs test does not suggest that either conference is superior.

$$\mu_G = \frac{2n_1n_2}{n_1 + n_2} + 1 = \frac{2 \cdot 26 \cdot 23}{26 + 23} + 1 = 25.40816$$

$$\sigma_G = \sqrt{\frac{2n_1n_2(2n_1n_2 - n_1 - n_2)}{(n_1 + n_2)^2(n_1 + n_2 - 1)}} = \sqrt{\frac{2 \cdot 26 \cdot 23(2 \cdot 26 \cdot 23 - 26 - 23)}{(26 + 23)^2(26 + 23 - 1)}} = 3.450091$$

$$z = \frac{G - \mu_G}{\sigma_G} = \frac{20 - 25.40816}{3.450091} = -1.57$$

11. The median is 2895.5; $n_1 = 25$; $n_2 = 25$; $G = 2$; $\mu_G = 26$; $\sigma_G = 3.49927$; Test statistic: $z = -6.86$; P-value $= 0.0000$; Critical values: $z = \pm 1.96$; Reject randomness. The sequence does not appear to be random when considering values above and below the median. There appears to be an upward trend, so the stock market appears to be a profitable investment for the long term.

$$\mu_G = \frac{2n_1n_2}{n_1 + n_2} + 1 = \frac{2 \cdot 25 \cdot 25}{25 + 25} + 1 = 26$$

$$\sigma_G = \sqrt{\frac{2n_1n_2(2n_1n_2 - n_1 - n_2)}{(n_1 + n_2)^2(n_1 + n_2 - 1)}} = \sqrt{\frac{2 \cdot 25 \cdot 25(2 \cdot 25 \cdot 25 - 25 - 25)}{(25 + 25)^2(25 + 25 - 1)}} = 3.49927$$

$$z = \frac{G - \mu_G}{\sigma_G} = \frac{2 - 26}{3.49927} = -6.86$$

13. a. List of sequences not provided.
    b. The 84 sequences yield these results: 2 sequences have 2 runs, 7 sequences have 3 runs, 20 sequences have 4 runs, 25 sequences have 5 runs, 20 sequences have 6 runs, and 10 sequences have 7 runs.
    c. With $P(2 \text{ runs}) = 2/84$, $P(3 \text{ runs}) = 7/84$, $P(4 \text{ runs}) = 20/84$, $P(5 \text{ runs}) = 25/84$, $P(6 \text{ runs}) = 20/84$, and $P(7 \text{ runs}) = 10/84$, each of the G values of 3, 4, 5, 6, 7 can easily occur by chance, whereas $G = 2$ is unlikely because $P(2 \text{ runs})$ is less than 0.025. The lower critical value of G is therefore 2, and there is no upper critical value that can be equaled or exceeded.
    d. The critical value of $G = 2$ agrees with Table A-10. The table lists 8 as the upper critical value, but it is impossible to get 8 runs using the given elements.

Quick Quiz

1. The ranks are 1, 3, 3, 5, 3. The rank for 7 is found using $\frac{2+3+4}{3} = 3$.
2. The efficiency rating of 0.91 indicates that with all other factors being the same, rank correlation requires 100 pairs of sample observations to achieve the same results as 91 pairs of observations with the parametric test for linear correlation, assuming that the stricter requirements for using linear correlation are met.
3. a. distribution-free test
   b. The term "distribution-free test" suggests correctly that the test does not require that a population must have a particular distribution, such as a normal distribution. The term "nonparametric test" incorrectly suggests that the test is not based on a parameter, but some nonparametric tests are based on the median, which is a parameter; the term "distribution-free test" is better because it does not make that incorrect suggestion.
4. Rank correlation should be used. The rank correlation test is used to investigate whether there is a correlation between foot length and height.
5. No, the P-values are almost always different, and the conclusions may or may not be the same.
6. Rank correlation can be used in a wider variety of circumstances than linear correlation. Rank correlation does not require a normal distribution for any population. Rank correlation can be used to detect some (not all) relationships that are not linear. 7. Because there are only two runs, all of the values below the mean occur at the beginning and all of the values above the mean occur at the end, or vice versa. This indicates the presence of an upward (or downward) trend.
7. Because there are only two runs, all of the values below the mean occur at the beginning and all of the values above the mean occur at the end, or vice versa. This indicates the presence of an upward (or downward) trend.
8. a. false
   b. false
9. Because the sign test uses only signs of differences while the Wilcoxon signed-ranks test uses ranks of the differences, the Wilcoxon signed-ranks test uses more information about the data and tends to yield conclusions that better reflect the true nature of the data.
10. Kruskal-Wallis test

Review Exercises

1. $r_s = 0.400$; The critical values are -0.700 and 0.700. Fail to reject the null hypothesis of $\rho_s = 0$. There is not sufficient evidence to support the claim of a correlation between job stress and annual income. Based on the given data, it does not appear that jobs with more stress have higher salaries.
2. Test statistic: $H = 2.5288$; (Tech: P-value $= 0.2824$); Critical value: $\chi^2 = 5.991$; Fail to reject the null hypothesis of equal medians. It appears that times of longevity after inauguration for presidents, popes, and British monarchs have the same median.

$$H = \frac{12}{N(N+1)}\left(\frac{R_1^2}{n_1} + \frac{R_2^2}{n_2} + \frac{R_3^2}{n_3}\right) - 3(N+1) = \frac{12}{76(76+1)}\left(\frac{1485.5^2}{38} + \frac{806.5^2}{24} + \frac{635^2}{14}\right) - 3(76+1)$$
$$= 2.5289$$

3. The test statistic of $z=\frac{(47+0.5)-\frac{110}{2}}{\sqrt{110}/2}=-1.43$ results in a P-value of 0.1527 and it is not less than or equal to the critical value of $z=-1.96$. Fail to reject the null hypothesis of $p=0.5$. There is not sufficient evidence to warrant rejection of the claim that in each World Series, the American League team has a 0.5 probability of winning.
4. $n_1=16$; $n_2=14$; $G=11$; critical values: 10, 22; Fail to reject randomness. There is not sufficient evidence to warrant rejection of the claim that odd and even digits occur in random order. The lottery appears to be working as it should.
5. The test statistic of $x=3$ is less than or equal to the critical value of 5 (from Table A-7). There is sufficient evidence to warrant rejection of the claim that the sample is from a population with a median equal to 5.
6. The test statistic $T=21$ is less than or equal to the critical value of 59. There is sufficient evidence to warrant rejection of the claim that the sample is from a population with a median equal to 5.
7. $R_1=204.5$; $R_2=230.5$; $\mu_R=255$; $\sigma_R=22.58318$; Test statistic: $z=-2.24$; (Tech: P-value $=0.025$); Critical values: $z=\pm1.96$; Reject the null hypothesis that the populations have the same median. There is sufficient evidence to warrant rejection of the claim that the recent eruptions and past eruptions have the same median time interval between eruptions. The conclusion does change with a 0.01 significance level.

$$\mu_R=\frac{n_1(n_1+n_2+1)}{2}=\frac{17(17+12+1)}{2}=255$$

$$\sigma_R=\sqrt{\frac{n_1n_2(n_1+n_2+1)}{12}}=\sqrt{\frac{17\cdot12(17+12+1)}{12}}=22.58318$$

$$z=\frac{Z-\mu_R}{\sigma_R}=\frac{204.5-255}{22.58318}=-2.24$$

8. The test statistic of $x=0$ is less than or equal to the critical value of 0. There is sufficient evidence to reject the claim of no difference. It appears that there is a difference in cost between flights scheduled 1 day in advance and those scheduled 30 days in advance. Because all of the flights scheduled 30 days in advance cost less than those scheduled 1 day in advance, it appears to be wise to schedule flights 30 days in advance.
9. The test statistic of $T=0$ is less than or equal to the critical value of 4. There is sufficient evidence to reject the claim that differences between fares for flights scheduled 1 day in advance and those scheduled 30 days in advance have a median equal to 0. Because all of the flights scheduled 30 days in advance cost less than those scheduled 1 day in advance, it appears to be wise to schedule flights 30 days in advance.
10. $r_s=0.714$. Critical values: $r_s=\frac{\pm z}{\sqrt{n-1}}=\frac{\pm1.96}{\sqrt{8-1}}=\pm0.741$; Fail to reject the null hypothesis of $\rho_s=0$. There is not sufficient evidence to support the claim that there is a correlation between the student ranks and the magazine ranks. When ranking colleges, students and the magazine do not appear to agree.

Cumulative Review Exercises

1. Flight 1: $\bar{x}=-1.3$ min; $Q_2=-2.0$ min; $s=1.6$ min

   Flight 19: $\bar{x}=9.9$ min; $Q_2=-0.5$ min; $s=26.6$ min

   Flight 21: $\bar{x}=33.4$ min; $Q_2=15.5$ min; $s=50.3$ min

   The means appear to be very different, with Flight 21 having the longest departure delay times. The medians appear to be very different, with Flight 21 having the longest departure delay times. The standard deviations appear to be very different, with Flight 21 having the greatest amount of variation. Flight 21 appears to be the least predictable flight because it has the highest variation, and it appears to have the longest departure delay times.

2. The normal quantile plot suggests that departure delay times for Flight 19 are not normally distributed.

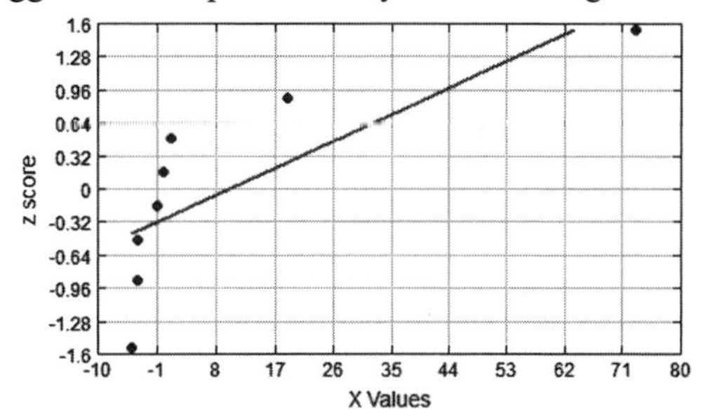

3. Kruskal-Wallis test statistic: $H = 3.2600$; Tech: P-value $= 0.1959$; Critical value $(\alpha = 0.05)$: $\chi^2 = 5.991$; Fail to reject the null hypothesis of equal medians. There is not sufficient evidence to warrant rejection of the claim that the three samples are from populations with the same median departure delay time.

$$H = \frac{12}{N(N+1)}\left(\frac{R_1^2}{n_1} + \frac{R_2^2}{n_2} + \frac{R_3^2}{n_3}\right) - 3(N+1) = \frac{12}{24(24+1)}\left(\frac{78^2}{8} + \frac{94^2}{8} + \frac{128^2}{8}\right) - 3(24+1)$$
$$= 3.2600$$

4. 95% CI: $\hat{p} \pm z_{\alpha/2}\sqrt{\frac{\hat{p}\hat{q}}{n}} = 0.039 \pm 1.96\sqrt{\frac{(0.039)(0.961)}{2000}} \Rightarrow 0.031 < p < 0.047$, or $3.1\% < p < 4.7\%$; We have 95% confidence that the limits of 3.1% and 4.7% actually contain the true percentage of the population of workers who test positive for drugs.

5. $H_0: p = 0.03$; $H_1: p > 0.03$; Test statistic: $z = \frac{0.039 - 0.03}{\sqrt{\frac{(0.03)(0.97)}{2000}}} = 2.36$;

   P-value $= P(z > 2.36) = 0.0091$ (Tech: 0.0092); Critical value: $z = 1.645$;

   Reject H_0. There is sufficient evidence to support the claim that the rate of positive drug test results among workers in the United States is greater than 3.0%.

6. The sample mean is 54.8 years. $n_1 = 19$; $n_2 = 19$; The number of runs is $G = 18$. The critical values are 13 and 27. Fail to reject the null hypothesis of randomness. There is not sufficient evidence to warrant rejection of the claim that the sequence of ages is random relative to values above and below the mean. The results do not suggest that there is an upward trend or a downward trend.

7. $n = \frac{[z_{\alpha/2}]^2 \hat{p}\hat{q}}{E^2} = \frac{[1.96]^2 \cdot 0.25}{0.02^2} = 2401$

8. There is a relatively small number of players with salaries that are substantially large, so the mean is strongly affected by those values, resulting in a large value of the mean, but the median is not affected by the small number of very large salaries.

9. $H_0: p = 0.5$; $H_1: p > 0.5$; Test statistic: $z = \frac{0.54 - 0.5}{\sqrt{\frac{(0.5)(0.5)}{285}}} = 1.36$;

   P-value $= P(z > 1.36) = 0.0869$ (Tech: 0.0865); Critical value: $z = 1.645$;

   Fail to reject H_0. There is not sufficient evidence to support the claim that the majority of the population is not afraid of heights in tall buildings. Because respondents themselves chose to reply, the sample is a voluntary response sample, not a random sample, so the results might not be valid.

10. There must be an error, because the rates of 13.7% and 10.6% are not possible with samples of size 100.

Chapter 14: Statistical Process Control

Section 14-1: Control Charts for Variation and Mean

1. No, if we know that the process is within statistical control, we know that none of the three out-of-control criteria are satisfied, but we know nothing about whether any specifications or requirements are satisfied. It is possible to be within statistical control while manufacturing altimeters with errors that are too large to satisfy the FAA requirements.

3. The mean is out of statistical control. The elevations have decreased substantially in recent years, so Lake Mead is becoming shallower. The decreases have been significant (and they are having a dramatic impact on the affected populations).

5. $\bar{\bar{x}} = 267.11 \text{ lb}$; $\bar{R} = 54.96 \text{ lb}$, $n = 7$

   For the R chart: $\text{LCL} = D_3\bar{R} = 0.076 \cdot 54.96 = 4.18 \text{ lb}$ and $\text{UCL} = D_4\bar{R} = 1.924 \cdot 54.96 = 105.74 \text{ lb}$.

   For the $\bar{x}$ chart: $\text{LCL} = \bar{\bar{x}} - A_2\bar{R} = 267.11 - 0.419 \cdot 54.96 = 244.08 \text{ lb}$ and

   $\text{UCL} = \bar{\bar{x}} + A_2\bar{R} = 267.11 + 0.419 \cdot 54.96 = 290.14 \text{ lb}$.

7. The R chart does not meet any of the three out-of-control criteria, so the variation of the process appears to be within statistical control.

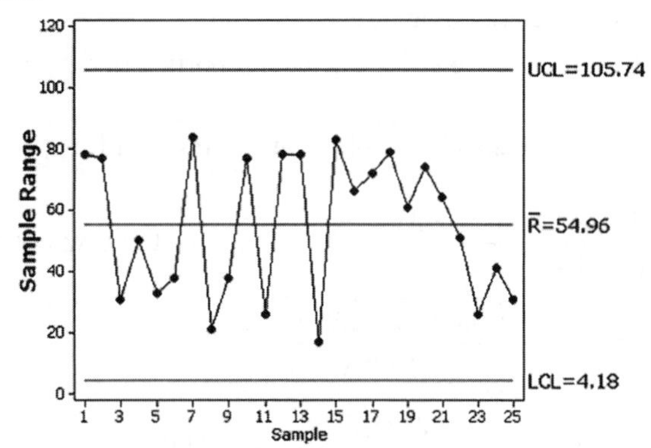

9. $\bar{\bar{x}} = 5.6955 \text{ g}$; $\bar{R} = 0.2054 \text{ g}$; $n = 5$

   For R chart: $\text{LCL} = D_3\bar{R} = 0.000 \cdot 0.2054 = 0.0000 \text{ g}$ and $\text{UCL} = D_4\bar{R} = 2.114 \cdot 0.2054 = 0.4342 \text{ g}$

   For $\bar{x}$ chart: $\text{LCL} = \bar{\bar{x}} - A_2\bar{R} = 5.6955 - 0.577 \cdot 0.2054 = 5.5770 \text{ g}$ and

   $\text{UCL} = \bar{\bar{x}} + A_2\bar{R} = 5.6955 + 0.577 \cdot 0.2054 = 5.8140 \text{ g}$

11. There are points lying beyond the upper control limit, so the process mean appears to be out of statistical control.

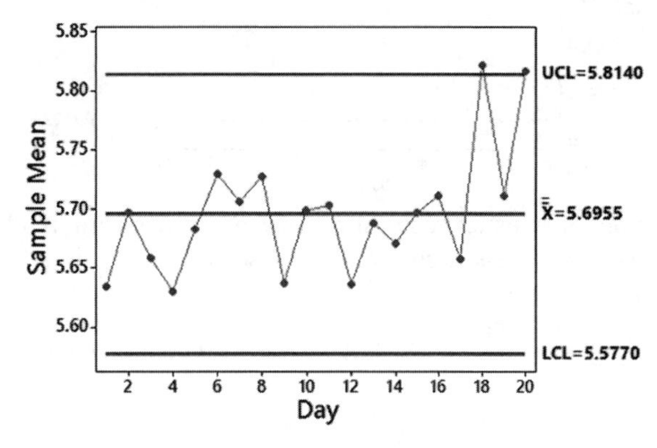

13. $\bar{s} = 12.27$ g; $n = 5$; $\text{LCL} = B_3\bar{s} = 0 \cdot 12.27 = 0$ g; $\text{UCL} = B_4\bar{s} = 2.089 \cdot 12.27 = 25.63$ g; Except for the values on the vertical scale, the s chart is nearly identical to the R chart shown in Example 3.

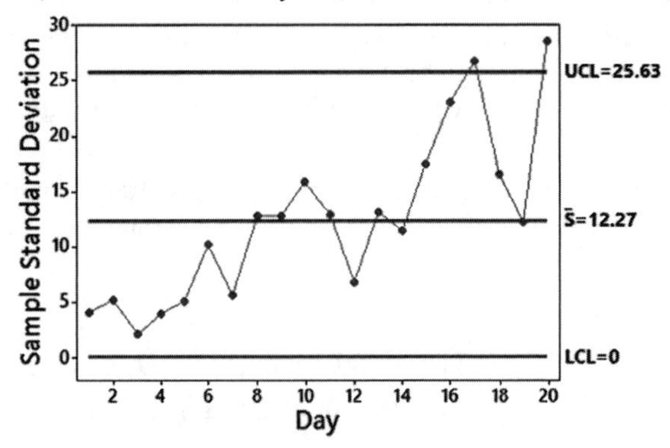

Section 14-2: Control Charts for Attributes

1. No, the process appears to be out of statistical control. There is a downward trend and there are at least eight consecutive points all lying above the centerline. Because the proportions of defects are decreasing, the manufacturing process is not deteriorating; it is improving.
3. Because the value of –0.00325 is negative and the actual proportion of defects cannot be less than 0, we should replace that value with 0.
5. The process appears to be within statistical control. (Considering a shift up, note that the first and last points are about the same.)

$$\bar{p} = \frac{32+21+25+19+35+34+27+30+26+33}{10{,}000 \cdot 10} = 0.00282;\ \bar{q} = 1 - 0.00282 = 0.99718$$

$$\text{LCL} = \bar{p} - 3\sqrt{\frac{\bar{p}\bar{q}}{n}} = 0.00282 - 3\sqrt{\frac{(0.00282)(0.99718)}{10{,}000}} = 0.001229$$

$$\text{UCL} = \bar{p} + 3\sqrt{\frac{\bar{p}\bar{q}}{n}} = 0.00282 + 3\sqrt{\frac{(0.00282)(0.99718)}{10{,}000}} = 0.004411$$

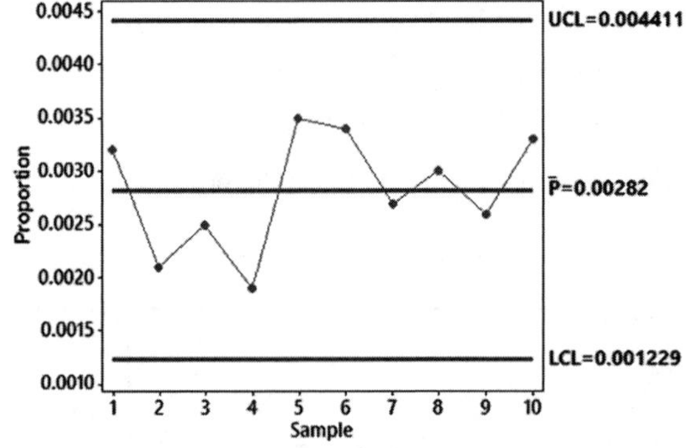

7. The process appears to be out of statistical control because of a downward trend, but the number of defects appears to be decreasing, so the process is improving. Causes for the declining number of defects should be identified so that they can be continued.

$$\bar{p} = \frac{16+18+13+9+10+8+6+5+5+3}{1000 \cdot 10} = 0.0093;\ \bar{q} = 1 - 0.0093 = 0.9907$$

$$\text{LCL} = \bar{p} - 3\sqrt{\frac{\bar{p}\bar{q}}{n}} = 0.0093 - 3\sqrt{\frac{(0.0093)(0.9907)}{1000}} = 0.00019$$

$$\text{UCL} = \bar{p} + 3\sqrt{\frac{\bar{p}\bar{q}}{n}} = 0.0093 + 3\sqrt{\frac{(0.0093)(0.9907)}{1000}} = 0.01841$$

7. (continued)

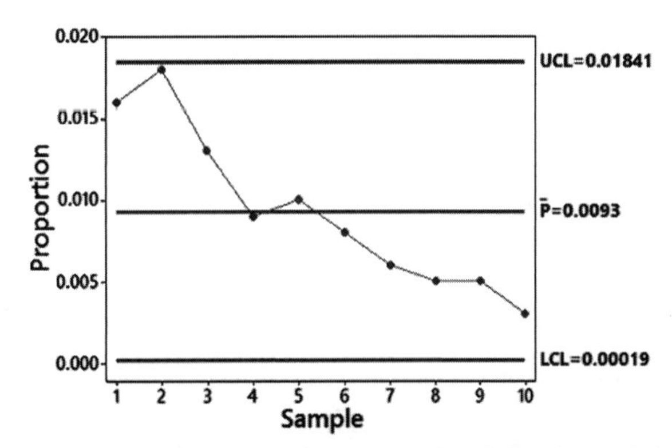

9. The process is out of statistical control because there are points lying beyond the upper control limit and there are points lying beyond the lower control limit. Also, there are eight consecutive points all lying below the centerline. The percentage of voters is increasing in recent presidential elections, and it should be much higher than any of the rates shown.

$$\bar{p} = \frac{631+619+608+552+536+526+531+501+551+491+513+553+568}{1000 \cdot 13} = 0.5523$$

$$\bar{q} = 1 - 0.5523 = 0.4477$$

$$\text{LCL} = \bar{p} - 3\sqrt{\frac{\bar{p}\bar{q}}{n}} = 0.5523 - 3\sqrt{\frac{(0.5523)(0.4477)}{1000}} = 0.5051$$

$$\text{UCL} = \bar{p} + 3\sqrt{\frac{\bar{p}\bar{q}}{n}} = 0.5523 + 3\sqrt{\frac{(0.5523)(0.4477)}{1000}} = 0.5995$$

11. Although the process is within statistical control, the proportions of defects are substantially high, so immediate corrective action should be taken to substantially lower the proportions of defects.

$$\bar{p} = \frac{8+7+9+8+10+6+5+7+9+12+9+6+8+7+9+8+11+10+9+7}{50 \cdot 20} = 0.165$$

$$\bar{q} = 1 - 0.165 = 0.835$$

$$\text{LCL} = \bar{p} - 3\sqrt{\frac{\bar{p}\bar{q}}{n}} = 0.165 - 3\sqrt{\frac{(0.165)(0.835)}{50}} = 0.0075$$

$$\text{UCL} = \bar{p} + 3\sqrt{\frac{\bar{p}\bar{q}}{n}} = 0.165 + 3\sqrt{\frac{(0.165)(0.835)}{50}} = 0.3225$$

11. (continued)

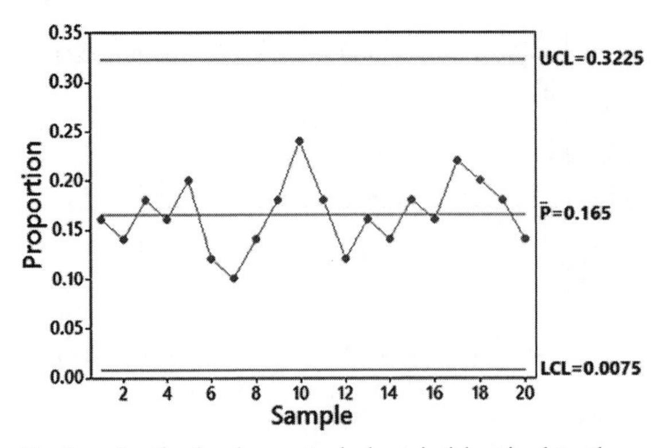

13. Except for a different vertical scale, the basic control chart is identical to the one given for Example 1.

$$\bar{p} = \frac{2+0+1+3+1+2+2+4+3+5+12+7}{100 \cdot 12} = 0.035;\ \bar{q} = 1 - 0.035 = 0.965;\ n\bar{p} = 100(0.035) = 3.5$$

$$\text{LCL} = n\bar{p} - 3\sqrt{n\bar{p}\bar{q}} = 3.5 - 3\sqrt{100(0.035)(0.965)} = -2.013, \text{ so LCL} = 0$$

$$\text{UCL} = n\bar{p} + 3\sqrt{n\bar{p}\bar{q}} = 3.5 - 3\sqrt{100(0.035)(0.965)} = 9.01$$

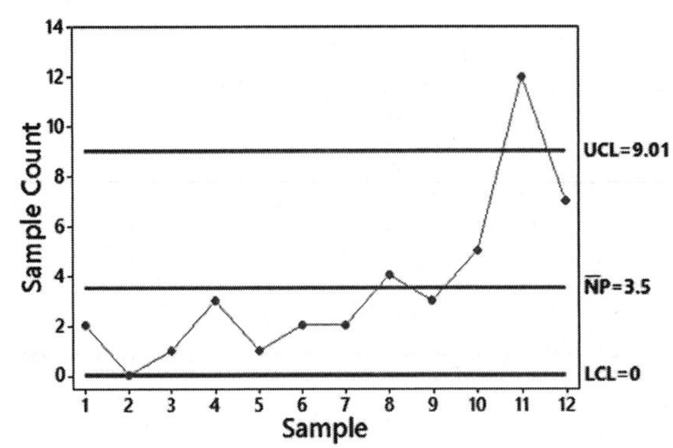

Quick Quiz

1. Process data are data arranged according to some time sequence. They are measurements of a characteristic of goods or services that result from some combination of equipment, people, materials, methods, and conditions.
2. Random variation is due to chance, but assignable variation results from causes that can be identified, such as defective machinery or untrained employees.
3. There is a pattern, trend, or cycle that is obviously not random. There is a point lying outside the region between the upper and lower control limits. There are at least eight consecutive points all above or all below the centerline.
4. An R chart uses ranges to monitor variation, but an $\bar{x}$ chart uses sample means to monitor the center (mean) of a process.
5. No, the R chart has at least eight consecutive points all lying below the centerline, there are at least eight consecutive points all lying above the centerline, there are points lying beyond the upper and lower control limits, and there is a pattern showing that the ranges have jumped in value for the more recent samples. What a mess!
6. $\bar{R} = 67.0$ ft; In general, a value of $\bar{R}$ is found by first finding the range for the values within each individual subgroup; the mean of those ranges is the value of $\bar{R}$.
7. No, the $\bar{x}$ chart has a point lying beyond the upper control limit, and there are at least eight consecutive points lying below the centerline.

8. $\bar{\bar{x}} = -2.24$ ft; In general, a value of $\bar{\bar{x}}$ is found by first finding the mean of the values within each individual subgroup; the mean of those subgroup means is the value of $\bar{\bar{x}}$.
9. No, the control charts can be used to determine whether the mean and variation are within statistical control, but they do not reveal anything about specifications or requirements.
10. Because there is a downward trend, the process is out of statistical control, but the rate of defects is decreasing, so we should investigate and identify the cause of that trend so that it can be continued.

Review Exercises

1. $\bar{\bar{x}} = 3157$ kWh; $\bar{R} = 1729$ kWh; $n = 6$

   For R chart: $\text{LCL} = D_3\bar{R} = 0.000 \cdot 1729 = 0$ kWh and $\text{UCL} = D_4\bar{R} = 2.004 \cdot 1729 = 3465$ kWh

   For $\bar{x}$ chart: $\text{LCL} = \bar{\bar{x}} - A_2\bar{R} = 3157 - 0.483 \cdot 1729 = 2322$ kWh and

   $\text{UCL} = \bar{\bar{x}} + A_2\bar{R} = 3157 + 0.483 \cdot 1729 = 3992$ kWh

2. The process variation is within statistical control.

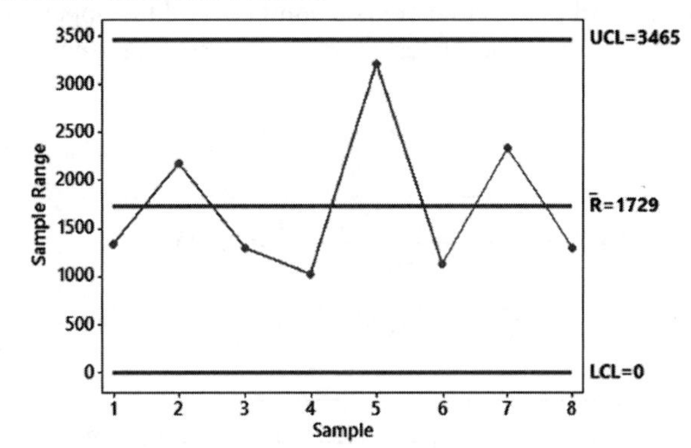

3. There appears to be a shift up in the mean values, so the process mean is out of statistical control.

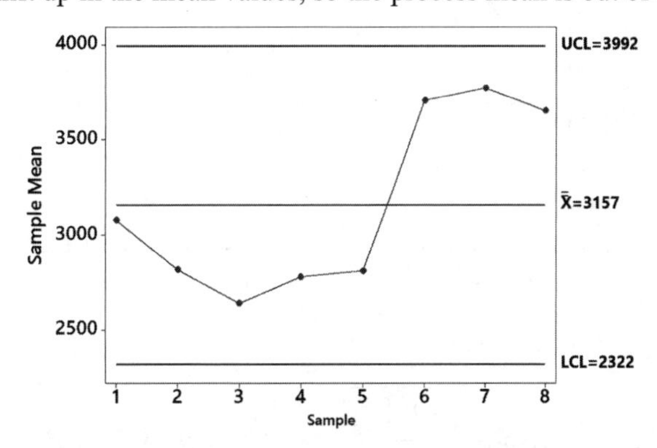

4. There appears to be a slight upward trend. There is 1 point that appears to be exceptionally low. (The author's power company made an error in recording and reporting the energy consumption for that time period.)

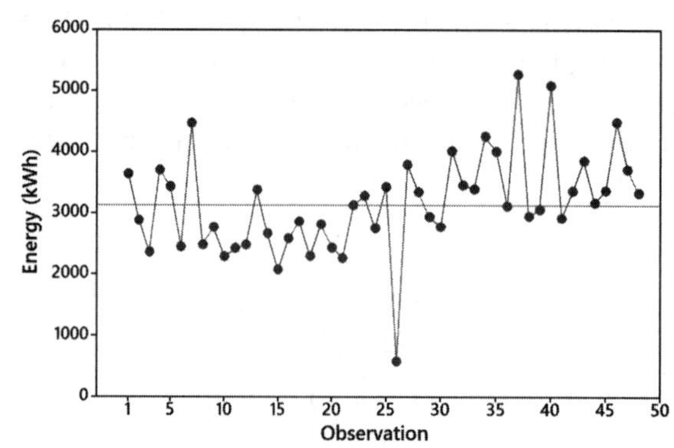

5. Because there is a distinct upward trend and there is a point beyond the upper control limit, the process is out of statistical control. Because the order times are clearly increasing, immediate corrective action should be taken.

$$\bar{p} = \frac{3+2+3+5+4+6+7+9+8+10+11+9+12+15+17}{50 \cdot 15} = 0.16133; \bar{q} = 1 - 0.16133 = 0.83867$$

$$\text{LCL} = \bar{p} - 3\sqrt{\frac{\bar{p}\bar{q}}{n}} = 0.16133 - 3\sqrt{\frac{(0.16133)(0.83867)}{50}} = 0.0053$$

$$\text{UCL} = \bar{p} + 3\sqrt{\frac{\bar{p}\bar{q}}{n}} = 0.16133 + 3\sqrt{\frac{(0.16133)(0.83867)}{50}} = 0.3174$$

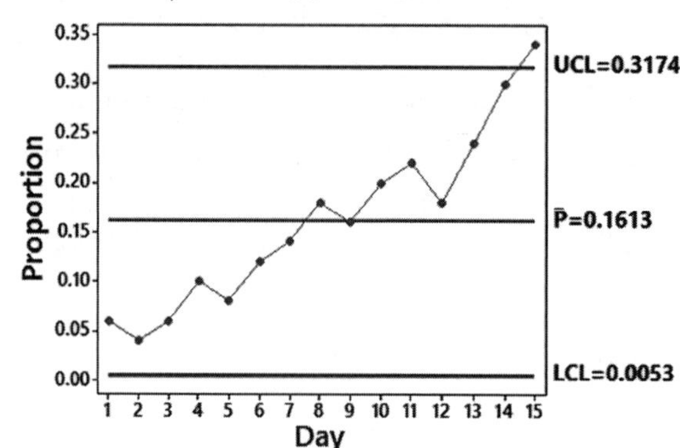

Cumulative Review Exercises

1. 95% CI: $\hat{p} \pm z_{\alpha/2}\sqrt{\frac{\hat{p}\hat{q}}{n}} = \frac{1108}{2015} \pm 1.96\sqrt{\frac{\left(\frac{1108}{2015}\right)\left(\frac{907}{2015}\right)}{2015}} \Rightarrow 0.528 < p < 0.573$; Because the entire range of values in the confidence interval includes values that are all greater than 0.5, it does appear that the majority of adults learn about medical symptoms more often from the Internet than from their doctor.

2. $H_0: p = 0.5; H_1: p > 0.5$; Test statistic: $z = \frac{\frac{1108}{2015} - 0.5}{\sqrt{\frac{(0.5)(0.5)}{2015}}} = 4.48$;

   P-value $= P(z > 4.48) = 0.0001$ (Tech: 0.0000); Critical value: $z = 1.645$; Reject H_0. There is sufficient evidence to support the claim that the majority of adults learn about medical symptoms more often from the Internet than from their doctor.

3. The graph is misleading. The vertical scale begins with a frequency of 800 instead of 0, so the difference between the "yes" and "no" responses is greatly exaggerated.

4. a. $(0.55)^3 = 0.166$ b. $1 - (0.45)^3 = 0.909$

5. $H_0: \rho = 0$; $H_1: \rho \neq 0$; $r = 0.0.356$; P-value $= 0.313$ (Table: P-value > 0.05); Critical values: $r \approx \pm 0.632$; Fail to reject H_0. There is not sufficient evidence to support a claim of a linear correlation between the DJIA and sunspot numbers. Because we do not expect any relationship between sunspot numbers and the behavior of stocks, this result is not surprising.
6. $\hat{y} = 9772 + 79.2x$; With no significant linear correlation, the best predicted value of the DJIA in the year 2004 is $\bar{y} = 13{,}423.6$, and that value is not close to the actual 2004 value of 10,855.

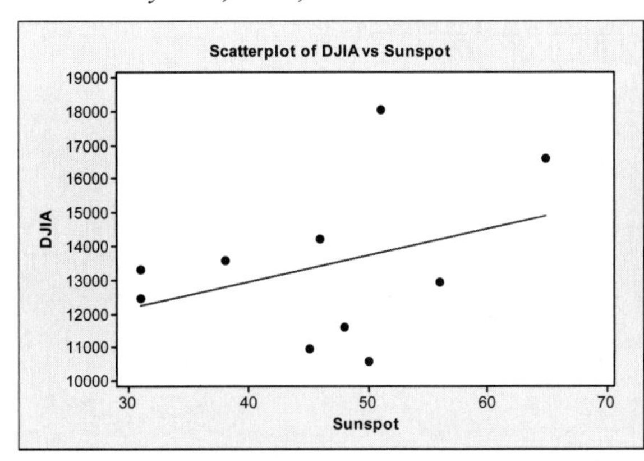

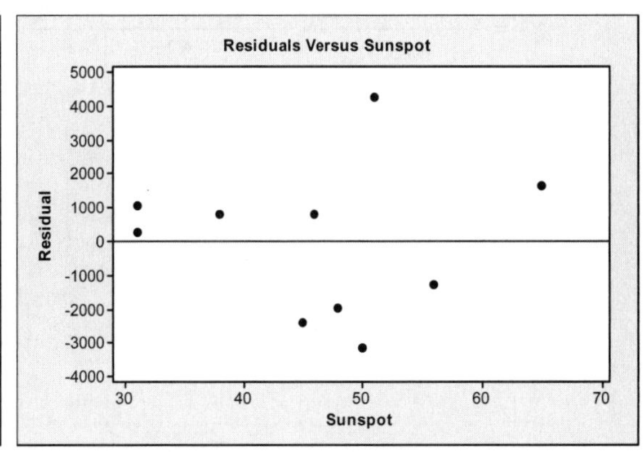

7. a. $z_{x=60} = \dfrac{60.0 - 68.6}{2.8} = -3.07$ and $z_{x=80} = \dfrac{80.0 - 68.6}{2.8} = 4.07$, which have a probability of $0.9999 - 0.0001 = 0.9998$, or 99.98% (Tech: 99.89%) between them.

   b. $z_{x=70} = \dfrac{70.0 - 68.6}{2.8/\sqrt{4}} = 1.00$; which has a probability of $1 - 0.8413 = 0.1587$ to the right.

8. There is a pattern of an upward trend, so the process is out of statistical control.

$$\bar{p} = \frac{3+2+4+6+5+9+7+10+12+15}{1200} = 0.060833;\ \bar{q} = 1 - 0.0608 = 0.93917$$

$$\text{LCL} = \bar{p} - 3\sqrt{\frac{\bar{p}\bar{q}}{n}} = 0.06083 - 3\sqrt{\frac{(0.06083)(0.93917)}{120}} = -0.0046, \text{ so LCL} = 0$$

$$\text{UCL} = \bar{p} + 3\sqrt{\frac{\bar{p}\bar{q}}{n}} = 0.06083 + 3\sqrt{\frac{(0.06083)(0.93917)}{120}} = 0.1263$$

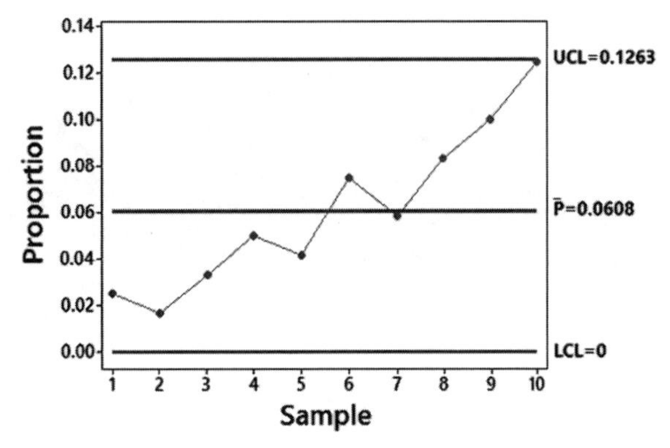

9. $\bar{x} = 7.3$; $Q_2 = 6.5$; $s = 4.2$; These statistics do not convey information about the changing pattern of the data over time.

10. H_0: Whether sentence is independent of plea.

    H_1: Whether sentence depends on plea.

    Test statistic: $\chi^2 = 42.557$; P-value $= 0.000$ (Table: P-value < 0.005); Critical value: $\chi^2 = 3.841$; df $= (2-1)(2-1) = 1$; Reject H_0. There is sufficient evidence to warrant rejection of the claim that the sentence is independent of the plea. The results encourage pleas for guilty defendants.

    $$\chi^2 = \frac{(392-418.48)^2}{418.48} + \frac{(58-31.52)^2}{31.52} + \frac{(564-537.52)^2}{537.52} + \frac{(14-40.48)^2}{40.48} = 42.557$$